IET POWER AND ENERGY SERIES 52

Nuclear Power

Other volumes in this series:

Volume 1	**Power circuit breaker theory and design** C.H. Flurscheim (Editor)
Volume 4	**Industrial microwave heating** A.C. Metaxas and R.J. Meredith
Volume 7	**Insulators for high voltages** J.S.T. Looms
Volume 8	**Variable frequency AC-motor drive systems** D. Finney
Volume 10	**SF6 switchgear** H.M. Ryan and G.R. Jones
Volume 11	**Conduction and induction heating** E.J. Davies
Volume 13	**Statistical technjiques for high voltage engineering** W. Hauschild and W. Mosch
Volume 14	**Uninterruptible power supplies** J. Platts and J.D. St Aubyn (Editors)
Volume 15	**Digital protection for power systems** A.T. Johns and S.K. Salman
Volume 16	**Electricity economics and planning** T.W. Berrie
Volume 18	**Vacuum switchgear** A. Greenwood
Volume 19	**Electrical safety: a guide to causes and prevention of hazards** J. Maxwell Adams
Volume 21	**Electricity distribution network design, 2nd edition** E. Lakervi and E.J. Holmes
Volume 22	**Artificial intelligence techniques in power systems** K. Warwick, A.O. Ekwue and R. Aggarwal (Editors)
Volume 24	**Power system commissioning and maintenance practice** K. Harker
Volume 25	**Engineers' handbook of industrial microwave heating** R.J. Meredith
Volume 26	**Small electric motors** H. Moczala *et al.*
Volume 27	**AC-DC power system analysis** J. Arrillaga and B.C. Smith
Volume 29	**High voltage direct current transmission, 2nd edition** J. Arrillaga
Volume 30	**Flexible AC Transmission Systems (FACTS)** Y-H. Song (Editor)
Volume 31	**Embedded generation** N. Jenkins *et al.*
Volume 32	**High voltage engineering and testing, 2nd edition** H.M. Ryan (Editor)
Volume 33	**Overvoltage protection of low-voltage systems, revised edition** P. Hasse
Volume 34	**The lighting flash** V. Cooray
Volume 35	**Control techniques drives and controls handbook** W. Drury (Editor)
Volume 36	**Voltage quality in electrical power systems** J. Schlabbach *et al.*
Volume 37	**Electrical steels for rotating machines** P. Beckley
Volume 38	**The electric car: development and future of battery, hybrid and fuel-cell cars** M. Westbrook
Volume 39	**Power systems of electromagnetic transients simulation** J. Arrillaga and N. Watson
Volume 40	**Advances in high voltage engineering** M. Haddad and D. Warne
Volume 41	**Electrical operation of electrostatic precipitators** K. Parker
Volume 43	**Thermal power plant simulation and control** D. Flynn
Volume 44	**Economic evaluation of projects in the electricity supply industry** H. Khatib
Volume 45	**Propulsion systems for hybrid vehicles** J. Miller
Volume 46	**Distribution switchgear** S. Stewart
Volume 47	**Protection of electricity distribution networks, 2nd edition** J. Gers and E. Holmes
Volume 48	**Wood pole overhead lines** B. Wareing
Volume 49	**Electric fuses, 3rd edition** A. Wright and G. Newbery
Volume 51	**Short circuit currents** J. Schlabbach
Volume 905	**Power system protection, 4 volumes**

Nuclear Power

J. Wood

The Institution of Engineering and Technology

Published by The Institution of Engineering and Technology, London, United Kingdom

First published 2007

The Institution of Engineering and Technology
Michael Faraday House
Six Hills Way, Stevenage
Herts, SG1 2AY, United Kingdom

www.theiet.org

British Library Cataloguing in Publication Data
Wood, Janet
Nuclear power. - (IEE power & energy series)
1. Nuclear reactors 2. Nuclear power plants
I. Title II. Institution of Engineering and Technology III. Institution of Electrical Engineers
621.4'83

ISBN (10 digit) 0 86341 668 3
ISBN (13 digit) 978-0-86341-668-2

Typeset in India by Newgen Imaging Systems (P) Ltd, Chennai
Printed in the UK by MPG Books Ltd, Bodmin, Cornwall

Contents

Chapter 1
The history of nuclear fission

1.1 Research and discovery

Uranium is the key to the exploitation of nuclear energy. As such, it has developed an exotic reputation, but it is in fact a very common element – as common as tin. It is present in rock and soil, and in water in trace amounts and, in much the same way as other minerals like tin, is found in various concentrations in different types of deposit. Rocks such as uraninite, autunite, uranophane, pitchblende or coffinite can be as much as 2 per cent uranium, but it also exists at a few parts per million in granite and many other rocks and at much greater concentration in deposits mined for uranium fuel.

Uranium oxide was used to give colour to ceramic glazes as far back as the first century AD, but it was not until the 18th century that it was isolated and named. It was separated from pitchblende (the so-called black mineral) in 1789, by a German chemist named Martin Klaproth, and named after the recently discovered planet Uranus, which in turn had been named after Urania, the muse of astronomy and geometry.

Uranium metal was first produced from uranium oxide by Eugene-Melchior Peligot in 1841. It is a silver-white metal, denser than lead.

The metal that had been so little known for so many centuries was the centre, over the next few decades, of a frenzy of scientific research and discovery that revealed that the atom had a complex internal structure and, moreover, one that was not unchangeable. Along the way, it also confirmed Albert Einstein's theories on mass and energy and laid the foundations for the exploitation of the power of the nucleus.

1.1.1 'Rays' from uranium

The first hint of effects that were later known to be caused by radioactivity was that uranium compounds were all found to spoil photographic plates by exposing them through their coverings, making them 'fogged'. The effect was initially assumed to be caused by x-rays, which had been discovered by William Roentgen in 1895. It was a French physicist, Henri Becquerel, who investigated the effects and was able to show that uranium's 'rays' were different: they could be deflected by an electric or magnetic field, which meant they carried an electric charge. Becquerel also found that the rays were a property of the uranium metal itself, and one that did not vary when it was in different chemical compounds.

In 1903 Becquerel won a Nobel Prize for his discovery, shared with two other French physicists, Marie and Pierre Curie. Following up Becquerel's observation of the fogging of photographic plates, Marie Curie observed that another substance, thorium, had a similar effect and named the effect 'radioactivity'. Observing that the radioactivity of pitchblende was greater than would be accounted for by the effect of uranium alone, she realised that it contained other substances with a similar property. She went on to isolate two more radioactive elements from the mineral: radium and polonium, which was named after her native Poland.

Pierre Curie died in 1906 but Marie Curie – who became the first female professor of physics at the Sorbonne – continued to investigate radioactivity, later working with her daughter Irene (also a Nobel Prize winner) and son-in-law Frederic Joliot. Marie Curie was a great advocate of the use of radioactivity: radium was used to treat malignant tumours (in a process then known as Curietherapy) and during the First World War she created mobile x-ray vans and trained operators to help the wounded.

These physicists who carried out the pioneering work on radioactivity have been remembered in the units used to measure the effects they discovered (see Panel 1.1). Meanwhile, radioactivity was being used both as a tool and as a marker to unlock the secrets of the atom's structure.

In England, New Zealander Ernest Rutherford had shown at the beginning of the century that in some cases elements emitted radioactive particles spontaneously. The radioactive emissions could be of more than one type, each of which had different properties, but in each case the emission left behind a different element.

In 1911 Frederick Soddy had discovered that naturally radioactive elements could take several different forms. While the forms were chemically identical, the weight of elements could differ. It became clear that the atom itself had a distinct structure and radioactivity represented a change in that structure. The source of these different versions was not discovered for nearly 20 years, when in 1932 James Chadwick discovered a heavy (in nuclear terms) particle with no charge which he called the neutron. Atoms of one element could contain a varying number of neutrons within their nucleus.

1.1.2 Inside the nucleus

The discoveries of electric charge and electricity over the 19th century had revealed that the electron could be separated from the rest of the atom. Indeed the electron appeared to be a relatively free agent, which could move from one atom to another or even exist as a static or moving electric charge. It had previously been proposed that electrons were embedded throughout an otherwise structureless atom 'like plums in

Panel 1.1 Measuring radioactive effect

The name of Pierre and Marie Curie, the scientists who discovered and explained radioactive processes, is recalled in the units used to express the amount and effect of radioactivity. New units have been developed that are consistent with SI (Système International) units but both versions may still be found in use.

The amount of radioactivity in a material is now expressed in becquerels (abbreviated Bq), where one Bq means one atom disintegrates (emits a radioactive particle) every second. The old unit for radioactivity was the Curie (abbreviation Ci), which is equal to 37 billion becquerels.

It is the energy of radiation which does the damage, and the amount of energy deposited in living tissue is referred to as the 'dose'. Doses are expressed in different ways depending on how much of the body, and what parts of it, are irradiated and the period during which the exposure takes place.

The amount of radiation energy that is absorbed per gramme of tissue is called the absorbed dose. It is measured in SI units called grays (abbreviated to Gy), where a gray corresponds to one joule of energy absorbed per kilogramme of matter. It was previously measured in rads, where a rad is equal to one hundredth of a gray.

The dose must be weighted for the potential of the radioactive particle to cause damage – alpha radiation is given twenty times the weight of gamma or beta radiation. This is known as the dose equivalent. There is another refinement to be made. Some parts of the body are more vulnerable than others, and a given dose of radiation is more likely to cause cancer in the lung, for example, than the thyroid. Different parts of the body are therefore also given weightings. Once this is done the dose is described as the effective dose equivalent, and in SI units it is expressed in sieverts (abbreviated Sv). The old measurement was called a rem, and one sievert is equal to 100 rem.

Exposure may also be measured in terms of the amount of charge absorbed. Once called the roentgen, this is now simply expressed in coulombs per kilogramme (C/kg).

a pudding', but the discovery of radioactivity changed that view. The whirl of new discoveries and theoretical work had, by the 1930s, given physicists a view into the atom that magnified it 10,000-fold.

By then it was clear that the atom was not 'solid' as had previously been assumed. Instead, it was made up of very light electrons orbiting a tiny, heavy nucleus. Inside the nucleus were densely packed particles, which were soon named protons and neutrons. The number of positively charged protons defined the element (and balanced the number of electrons in an uncharged atom). The number of neutrons, with no charge and a similar mass to that of the proton, could vary.

However, physicists found that the variations were not entirely random, because many proton–neutron configurations were not stable. Sooner or later the unstable configurations emitted a radioactive particle and settled into a new configuration – a process that could happen several times until a stable nucleus was finally achieved.

Physicists now had two numbers to define elements. The 'atomic number' is the number of protons in the nucleus. The unique number defines the element and corresponds to its place on the periodic table. A second number, the atomic mass, is the number of protons plus neutrons in the nucleus and it can vary. Carbon, for example, always has six protons (atomic number 6) but its atomic mass is most likely to be 12, 13 or 14 depending on whether it has six, seven or eight neutrons (and there are still more versions that are uncommon). The two lighter isotopes are stable but carbon-14 is not and undergoes radioactive decay (a property that has been used in archaeology in radiocarbon dating).

Atoms are now conventionally represented with their mass and number as follows: ${}^{12}_{6}C$, ${}^{13}_{6}C$ or ${}^{14}_{6}C$ or less formally as carbon-14, carbon-12, etc.

1.1.3 Radioactive decay

In observing elements which were radioactive, physicists had discovered three types of radioactive decay.

- The alpha (α) particle is made up of two protons and two neutrons – identical to the nucleus of the second-lightest element, helium. An alpha emission leaves behind a nucleus with an atomic weight decreased by four and an atomic number decreased by two. Uranium-238, for example, emits an alpha particle to form Thorium-234.
- The beta particle (β) is a high-energy electron. Physicists later concluded that it is produced when a neutron splits, emitting an electron and leaving behind a positively charged proton and an atomic nucleus with an atomic number that has increased by one. Carbon-14 (with six protons and eight neutrons), for example, emits a beta particle to form nitrogen-14, with seven protons and seven neutrons.
- Gamma radiation (γ) is an electromagnetic wave that is often emitted along with an alpha or beta particle.

Almost immediately after radioactivity was discovered, physicists began to use it as both an experimental tool and a subject. Taking particle-emitting substances they directed the emissions at 'target' elements. Rutherford, for example, used an alpha emitter to bombard nitrogen. They found that collisions between the radioactive particles and nuclei in the target sometimes struck radioactive particles from the target nucleus. In some cases, the nucleus absorbed the colliding particle. Rutherford found that a nitrogen target bombarded with beta particles in some cases absorbed the alpha particle to form oxygen (with eight protons) and hydrogen.

1.1.4 Absorption and fission

Later work by Irene Curie and Frederic Joliot, and by Enrico Fermi, found that a number of transformations could be caused by alpha and neutron particles. They used

Panel 1.2 Radioactivity and its effects

The three most common forms of radiation are described as alpha, beta and gamma. They have very different properties: they are different in form and are emitted with different energies and different penetrating power, so they present very different hazards to human beings.

- An alpha particle contains two neutrons and two protons. It is several thousand times heavier than a beta particle and has a double electric charge. It has almost no penetrating power. It cannot pass through a sheet of paper and therefore is unable to penetrate the skin's outer layers. It is not dangerous (except at very high activity levels, when it may cause burns) unless substances emitting it get into the body through an open wound or are eaten or breathed in, when it is much more damaging than either beta or gamma radiation.
- Beta radiation is effectively a very high energy electron, so it is much lighter than an alpha particle and carries a single (negative) charge. It is much more penetrating than an alpha particle and may penetrate living tissue to a depth of one or two centimetres. It can be blocked by metal sheets.
- Gamma radiation is not a particle but an electromagnetic wave similar to – but carrying more energy than – x-rays. As such it has very high penetrating power and shielding made of thick lead or concrete is required. There is an overlap between the energy ranges of x-rays and gammas. The fundamental difference is where they come from: x-rays are from the electron shells of the atom, whereas gammas are from the nucleus.

heavy elements like uranium as a target and produced both new elements and isotopes of existing ones.

Most of the results were in the predicted range: the nucleus was apparently absorbing neutrons or protons and the result was a different isotope or an element with an atomic number differing by one or two from the original target. The results varied: some transformations could produce wholly new elements, with a higher proton count in the nucleus than was known in nature, although such elements tended to be unstable and short-lived. That was a discovery for which Curie and Joliot won a Nobel Prize.

In most cases the resulting elements were isotopes of uranium or neighbouring elements. But some collisions were different and resulted in much lighter elements. This was very unexpected, but eventually radiochemist Otto Hahn identified the lighter products. He concluded that some collisions had produced isotopes of barium, which has 56 protons and a nucleus almost half the size of uranium's, which has an atomic number of 92. This was a nuclear disintegration very different from the loss of a radioactive particle that was by then familiar.

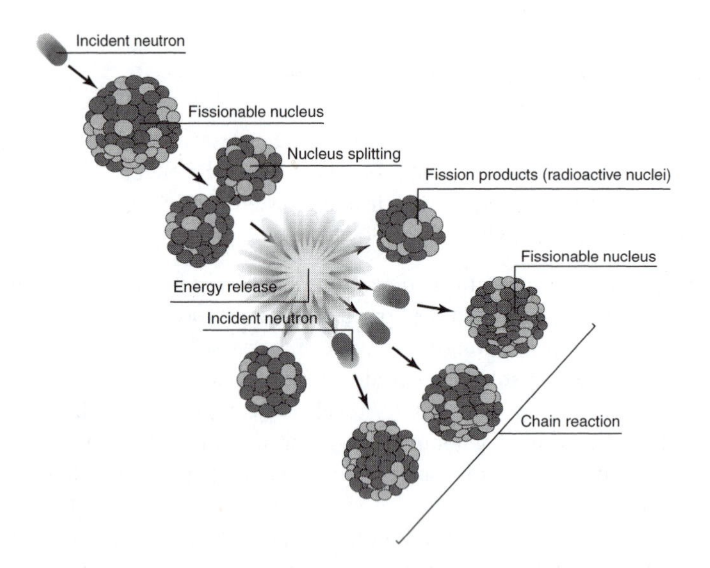

Figure 1.1 *In nuclear fission the uranium nucleus first absorbs a neutron and then splits. As it does so it emits neutrons that can initiate the process in another nucleus.*

The process at work was explained by Lise Meitner and Otto Frisch in 1939. They suggested that the nucleus had 'captured' a neutron as had been observed in other transformations. But the result had been a nucleus so unstable that it soon split into two – a process referred to as fission. The uranium nucleus had split into barium (atomic number 56) and krypton (atomic number 36), which had not been observed because it is a gas at room temperature and had escaped the experimental equipment.

The fission released a huge amount of energy – 200 million electron volts. Almost incidentally, this was the first experimental confirmation of Albert Einstein's 1905 paper that had expressed the relationship between mass and energy as $E = mc^2$. But for many physicists fission – if it could be carried out on a scale that allowed the energy to be used – raised the possibility of using nuclear energy to produce power or even to develop an atomic bomb.

1.2 Exploiting the fission process

The uranium fission process was slightly more complicated than it had first appeared.

First, it was already known that naturally occurring uranium is a mixture of at least three isotopes. By far the largest proportion – around 99.27 per cent – is uranium-238, which has 146 neutrons, along with its 92 protons. Some 0.72 per cent of the naturally occurring mineral is ^{235}U, which has just 143 neutrons. There is an even smaller proportion (0.0055 per cent) of ^{234}U, with one neutron fewer.

Both ^{235}U and ^{238}U can undergo fission, but it is more likely in ^{235}U and the results of the split are somewhat different because the energy required to cause fission varies.

When a uranium nucleus absorbs a neutron and splits into barium and krypton, some of the neutrons are not captured by the new elements but instead are released as 'free' neutrons. These free neutrons do not remain at large: instead they are in turn captured by the nuclei of nearby uranium atoms, where they can initiate another fission.

When ^{238}U is subject to fission it may or may not produce a free neutron, so on average there is less than one free neutron for each fission, and the process gradually dies away. In contrast, ^{235}U releases several free neutrons when it undergoes fission – 2.5 on average. This means that if there is enough ^{235}U in the uranium mix the fission reaction can be 'self-sustaining' so that at least one of the neutrons from each fission finds another ^{235}U nucleus and is absorbed to initiate another fission in a 'chain reaction'.

1.2.1 The chain reaction

How many of the neutrons initiate fission – and whether a chain reaction can be started – depends partly on the proportion of ^{235}U in the mix and partly on the total volume of material. The amount of fissile material needed for a sustained nuclear chain reaction is called the critical mass. It varies depending upon the nuclear and physical properties of the material. The important physical properties include the density and shape of the mass, as well as its purity. The nuclear properties include the nucleus's ability to capture a neutron (known as the nuclear fission cross section and generally fixed for each nucleus) and whether the process is aided by a neutron 'reflector' (which would send the neutron back into the mass) or 'moderator' (which would slow the neutron to a speed that makes it easily captured) or is interrupted by an 'absorber' (a material that removes neutrons from the process).

An assembly in which a chain reaction is just possible is called critical, and when the reaction becomes self-sustaining it is said to have reached criticality. In such an assembly, without a new input of free neutrons, for example from spontaneous fissions, the reaction will on average be just sustained.

At 2–3 per cent uranium-235 the reaction can be simply self-sustaining. This is believed to have happened in nature, at the Oklo deposits in Gabon, where a high uranium concentration was combined with other conditions to allow a self-sustaining reaction to take place over many years. The reaction is now over, but evidence remains in the proportions of elements within the rock deposit.

If an assembly is less than critical, the fission reaction will reach a steady state only with a steady input of new free neutrons, and the assembly is said to be subcritical.

A more than critical assembly is said to be supercritical. An assembly that is capable of sustaining a chain reaction without needing the contribution of defined neutrons is called prompt critical (and is therefore also supercritical). Even larger masses are called superprompt critical.

If ^{235}U makes up most, or all, of the sample, the chain reaction may be explosive.

1.2.2 Making plutonium

Along with energy and the fission products barium and krypton, the bombardment of uranium with neutrons also results in other products, from smaller disintegrations and neutron capture that is not followed by fission. Among the most important is plutonium.

Plutonium, one of the two fissile elements used to fuel nuclear explosives, is not found in significant quantities in nature. It has an atomic number of 94, meaning its nucleus contains two more protons than uranium and it must be produced from uranium-238.

In this route uranium-238 captures a neutron to become uranium-239, which beta decays within a few hours to become neptunium-239. The neptunium isotope again emits a beta particle (this time over a timescale of several days) to become plutonium-239. Like uranium, but unlike the two intermediate substances, plutonium is extremely slow to undergo radioactive decay in nature. Its five different isotopes take between several and many thousand years to decay. Radiologically, it is relatively safe to handle because it emits alpha particles (although ^{241}Pu is principally a beta emitter) which can be blocked by simple shielding (see Panel 1.2), although they are hazardous at very short distances or inside the body.

The importance of plutonium is that when it captures a neutron, plutonium-239 undergoes fission more readily than uranium, and in the process it produces more excess neutrons to continue the reaction. This means that a smaller mass can reach criticality, with a self-sustaining or explosive chain reaction. This can make chemical handling of plutonium difficult, as volumes – of the solid and its various compounds, in solution or liquid form – have to be kept small so that criticality does not occur.

In generating power from the nuclear reaction, plutonium-239 is another fissile isotope, albeit one that requires careful handling from both a chemical and criticality point of view.

1.2.3 Controlling the reaction

The proportion of ^{235}U is not the only factor that determines whether and how fast the chain reaction takes place.

The high-energy neutrons emitted during fission can cause more conventional radioactive decay, knocking out a radioactive particle and leaving a heavy element but not initiating fission. It was Niels Bohr who first suggested that the speed of the neutron was an important factor in the fission process, and Enrico Fermi, working in the US, who developed the theory of a 'moderator' – a substance that could slow down high-energy neutrons in a series of collisions until they were moving slowly enough to be captured by another uranium nucleus. Moderators could include water or graphite.

Fermi's work marked a decisive shift from nuclear theory and experiment to practical exploitation of the nuclear reaction. There was a dual emphasis on the programme at this time. It was clear that fission energy could potentially be used for power generation, but this was 1940, and the potential for a weapon of devastating – and decisive – power took precedence. The Manhattan Project to develop that weapon is outside the scope of this book, but the techniques and technologies developed to fuel, manage and exploit the nuclear reaction are an important part of the story of nuclear power.

1.2.4 Fermi and the Chicago Pile

Until 1942 nuclear fission had remained a laboratory-scale experimental process. But at the end of that year the first nuclear reactor, Chicago Pile 1 (CP-1), went critical for the first time in a racquet court at the University of Chicago, under the direction of Enrico Fermi.

Born in Italy, Fermi had been working on radioactivity and nuclear physics since the early 1930s at the University of Rome and in 1938 he won the Nobel Prize 'for his demonstrations of the existence of new radioactive elements produced by neutron irradiation, and for his related discovery of nuclear reactions brought about by slow neutrons'. In that year Fermi also emigrated to the USA. He became professor of Physics at the University of Columbia and later a leading physicist on the Manhattan Project.

CP-1 was known as the 'pile' because it was built of layers of uranium fuel lumps and bricks of graphite. The completed pile was nearly 8 m wide and 6 m high and it contained nearly 350,000 kg of graphite, 36.5 kg of uranium oxide and over 5,500 kg of uranium metal when it went critical.

It took just a month from when work started on building CP-1 to reach the point of achieving criticality. The pile itself produced only half a watt of power – barely enough to light a small torch. But it demonstrated both the feasibility of the reaction and the solution of the most important problems involved in exploiting nuclear power: how to start the reaction, how to control it and how to halt it at will. Their story is told by both the National Museum of Nuclear Science and History and the University of Chicago's high-energy physics historical resource.

1.2.5 Building the pile

Fermi, working with Walter H Zinn and a team of other scientists and technologists, had to determine operationally possible designs of a uranium chain reactor.

First, they had to find a suitable 'moderator' that would slow the fast neutrons which escaped during the reaction so that they could be captured by another fissile nucleus. What was needed was a material that could be used to get a 'reproduction factor' greater than one.

At that time it was not clear that a reproduction factor ('k') which would allow the chain reaction to sustain itself would be achievable. Experimenters did not even know whether a k value bigger than one was physically possible or whether a k value could be achieved which was high enough to balance other factors that would be slowing down the reaction. Those factors might include impurities in the uranium and in the moderator that would capture neutrons and take them out of the fission process. Another factor could be neutrons escaping outside the pile without encountering uranium-235 atoms. From the last point it is clear that the size of the pile would be one factor in determining whether a chain reaction could be obtained, and it was entirely possible that the necessary size of the pile would be so great that criticality would remain a theoretical goal but not practically achievable.

According to Atomic Archive, one of the first things that had to be determined was how best to place the uranium in the reactor. A cubical lattice of uranium in a matrix of the moderating material appeared to offer the best opportunity for a neutron to encounter a uranium atom. Graphite was chosen as the moderator because it could be obtained at the necessary purity. As for the uranium fuel, it was necessary to use uranium oxides because metallic uranium of the desired degree of purity did not exist.

Graphite–uranium lattice reactors had first been studied at Columbia in 1941, before the nuclear programme was brought together at the University of Chicago. Once at Chicago, researchers worked on a series of small subcritical piles that would provide the data needed to design CP-1 – eventually working through 30 different piles. By July 1942, the measurements obtained from these experimental piles had gone far enough to permit a choice of design for a test pile of critical size, and researchers could decide on the shape and size of the uranium oxide fuel elements.

When the time came to construct CP-1 it was enclosed in a balloon cloth bag which could be evacuated to remove air from the pile (as air would have a neutron-capturing effect and slow the chain reaction). In the centre of the floor a circular layer of graphite bricks was placed. This, and each succeeding, layer of the pile was braced by a wooden frame. Alternate layers contained the uranium. By this layer-on-layer construction a roughly spherical pile of uranium and graphite was formed.

Before the structure was half complete, measurements indicated that the critical size at which the pile would become self-sustaining was somewhat less than had been anticipated in the design.

The pile consisted of uranium pellets as a neutron-producing ‘core’, separated from one another by graphite blocks to slow the neutrons. Fermi himself described the apparatus as ‘a crude pile of black bricks and wooden timbers’.

1.2.6 Controlling the reaction

With limited knowledge about the chain reaction that was to be produced in CP-1, control of the reaction was vital.

The key to controlling the speed of the reaction was including a material that would absorb neutrons in the pile but organising it in such a way that the ‘absorber’ could be inserted or withdrawn at will. When the absorber was gradually withdrawn, the number of free neutrons would rise to a level that would initiate and maintain fission and promote the reaction, and when reinserted it would reduce the reaction again.

Fermi chose cadmium as the absorber, machined into three rods that reached through the pile. Withdrawing the rods would increase neutron activity in the pile to lead to a self-sustaining chain reaction. Reinserting the rods would dampen the reaction.

The first of the three rods was operated by hand. The second, driven by a motor, was to be triggered automatically when core reactivity reached a certain level.

The third rod was an emergency system. It was held in place above the reactor by a rope and pulley and it depended on the reflexes of a man with an axe, who was standing by. If necessary, the man would chop the rope so the control rod would drop into the reactor and shut down the reaction before it could get out of control. This 'safety control rod axe man' is said to be the origin of the now-common acronym 'scram', used for an emergency shutdown of a reactor.

On 2 December 1942 the pile was allowed to go critical – measured by a sudden increase in the neutrons being emitted within the pile and an increase in temperature in the graphite blocks.

In the Manhattan Project, the aim of the reactor was to produce plutonium that could be used to make a bomb. The heat in the pile from the reaction was a by-product. But scientists were already working on designs to use the heat from the fission reaction to produce electricity.

1.2.7 The British Programme

The Manhattan Project absorbed scientists and engineers from across the USA as well as many arriving from Europe, but the work on discovering the structure of the atom, fission and its potential uses had been an international one. In Russia, for example, there were several research centres specialising in nuclear physics by the 1930s.

In Britain, two refugee physicists, Peierls and Frisch, working at Birmingham University, had given a major impetus to the effort in a three-page document known as the Frisch–Peierls Memorandum written in 1940. They predicted that about 5 kg of pure ^{235}U could make a very powerful atomic bomb equivalent to several thousand tonnes of dynamite. This memorandum stimulated a considerable response in Britain.

A group of scientists known as the MAUD Committee was set up and supervised research at five universities (Birmingham, Bristol, Cambridge, Liverpool and Oxford). The chemical problems of producing gaseous compounds of uranium and pure uranium metal were studied at Birmingham University and Imperial Chemical Industries.

It was the work at Cambridge that provided the first experimental proof that a chain reaction could be sustained with slow neutrons in a mixture of uranium oxide and heavy water. The Cambridge team also first discovered that the probability of fission in ^{235}U is much greater than in ^{238}U when it absorbs a slow neutron, whereas the ^{238}U is more likely to form ^{239}U and then undergo two nuclear disintegrations to form new elements. The work was confirmed with parallel experiments in the USA and both teams agreed on the names neptunium for the new element with atomic number 93 and plutonium for the element with atomic number 94. The identification of plutonium in 1941 is generally credited to Glenn Seaborg.

By the end of 1940 remarkable progress had been made by the several groups of scientists coordinated by the MAUD Committee. The final outcome of the MAUD Committee, just 15 months after it was set up, was two summary reports in July 1941. One was on the 'Use of Uranium for a Bomb' and the other was on the 'Use of Uranium as a Source of Power'.

The first report concluded that a bomb was feasible and that one containing some 12 kg of active material would be an equivalent of 1,800 tons of TNT and would release large quantities of radioactive substances which would make places near the explosion site dangerous to humans for a long period.

The second MAUD Report concluded that the controlled fission of uranium could be used to provide energy in the form of heat for use in machines, as well as providing large quantities of radioisotopes which could be used as substitutes for radium in applications such as medicine. It referred to the use of heavy water and possibly graphite as moderators for the fast neutrons and said that even ordinary water could be used if the uranium was enriched in the ^{235}U isotope. It concluded that the 'uranium boiler' had considerable promise for future peaceful uses but that it was not worth considering during the war.

1.3 After the war

After the war thoughts quickly turned to a nuclear future and there were immediate fears over unrestricted access to nuclear technology – whether for bombs or for power. An Agreed Declaration in November 1945 by the US, Britain and Canada called for international control of nuclear energy. President Truman later said 'The hope of civilization lies in international arrangements looking, if possible, to the renunciation of the use and development of the atomic bomb, and directing and encouraging the use of atomic energy and all future scientific information toward peaceful and humanitarian ends'.

In July 1946 the United States passed an Atomic Energy Act, establishing the Atomic Energy Commission (AEC). The AEC replaced the Manhattan Project on 31 December 1946. During the first half of 1946, Congress debated whether atomic energy should be under civilian or military control and in the end an Act of Congress placed further development of nuclear technology under civilian rather than military control. Senator McMahon, the author of the Act (also known as the McMahon Act), called it 'a radical piece of legislation' because it gave the AEC a monopoly over both military and commercial uses of atomic energy. The Act said that atomic energy should be directed 'toward improving public welfare, increasing the standard of living, strengthening free competition among private enterprises . . . and cementing world peace'. However, the Act prohibited private companies or individuals from owning nuclear materials and patenting inventions related to atomic energy. The Act also restricted information on using nuclear materials to produce energy as well as on designing, making and using atomic weapons. The restrictions made the development of nuclear power much more difficult and in 1954 a new Atomic Energy Act was passed, intended to promote the peaceful uses of nuclear energy through private enterprise and to implement President Eisenhower's Atoms for Peace Program.

The Act allowed the Atomic Energy Commission to license private companies to use nuclear materials and build and operate nuclear power plants. This Act amended the Atomic Energy Act of 1946, which had placed complete power of atomic energy development in the hands of the Atomic Energy Commission. Parallel work on developing nuclear power technologies was under way in at least three countries.

In 1947 the US Congress authorised work on a nuclear submarine and a nuclear-powered aeroplane. The aeroplane option was never pursued – not least because the weight of shielding required would be prohibitive. However, the submarine option was ripe for exploitation, and the possibility of using a nuclear propulsion unit had initially been one of the US's major lines of nuclear research. Submarines up to the end of the war were diesel-powered or battery-powered, so their range and speed were severely restricted. The need to refuel, run engines to charge their batteries or refresh the air on board compromised their ability to stay submerged. A nuclear-powered submarine, in comparison, could stay submerged indefinitely, and certainly for an entire voyage. The huge increase in power available also meant a nuclear-powered submarine could sustain high speeds for the whole trip.

The potential benefits of a nuclear-powered submarine drove a very fast development programme, the results of which are still being felt today as it became the basis for the pressurised-water reactor that is used around the world for power generation. The pressurised-water reactors were originally designed for use in nuclear submarines by Westinghouse under the direction of Admiral Hyman C Rickover. The needs of the submarine environment largely dictated the PWR's design: a very compact core, with sealed circuits for transferring the energy from the core to the power unit, and one where fuel remained in the reactor for a year or more. Since they had to operate for long periods in a sealed, submerged vessel, these reactors had to be designed with a minimum of radioactive leakage either into the submarine, where the crew had to live for months at a time, or into the water, where the bubbles of radioactive gases would permit easy detection of the submarine's position.

1.3.1 Early reactor designs

A design for a pressurised-water submarine thermal reactor was first developed by the Argonne Laboratory for Westinghouse in 1947. The Argonne team designed and developed the reactor core for the world's first atomic-powered submarine and, in 1950, built and operated the first submarine reactor prototype, the Zero Power Reactor 1 (ZPR-1), launching the first atomic submarine, the *USS Nautilus*, in January 1954. The Argonne-designed reactor in the *Nautilus* lasted 62,500 miles including a dramatic crossing of the Arctic Ocean in 1958. Its scientific mission determined that the ocean depth at the North Pole, two-and-a-half miles, was far greater than previously estimated.

On 20 December 1951, the first usable electricity from nuclear energy was produced at the National Reactor Testing Station, later called the Idaho National Engineering Laboratory (INEL), in Idaho Falls, Idaho. The electricity lit four light bulbs strung across a railing in the turbine room of the Experimental Breeder Reactor I (EBR-I). The first reactor project approved by the Atomic Energy Commission, EBR-I was the brainchild of Walter Zinn, who had been a prime mover in the Chicago Pile project and was now heading the Argonne Laboratory. In 1953, EBR-I scientists showed a reactor could create more fuel than it used, that is, the reactor could 'breed' fuel as it created electricity. EBR-I operated as a research reactor until 1963.

The Russian approach had been to modify existing graphite-moderated channel-type plutonium production reactors for heat and electricity generation, and in 1954 the world's first nuclear-powered electricity generator began operation in the then closed city of Obninsk at the Institute of Physics and Power Engineering (FEI). The AM-1

(Atom Mirny – peaceful atom) reactor was water-cooled and graphite-moderated, with a design capacity of 30 MWt or 5 MWe. It was similar in principle to the plutonium production reactors in the closed military cities and served as a prototype for other graphite channel reactor designs including the Chernobyl-type RBMK (Reaktor Bolshoi Moshchnosty Kanalny – high-power channel reactor) reactors. AM-1 produced electricity until 1959 and was used until 2000 as a research facility and for the production of isotopes.

The UK also decided to use the graphite-moderated air-cooled approach, following its work during the war at Windscale, which was the site chosen to build two production reactors known as the Windscale Piles. Responsibility for the programme initially lay with the Ministry of Supply, which up to 1954 oversaw construction and operation of the Piles, an Atomic Energy Research Establishment at Harwell in Oxfordshire and uranium production facilities at Springfields near Preston in north-east England. The Ministry of Supply set up the construction of the UK's first nuclear power plant, also on the Windscale site, but by the time it started up a new organisation, the UK Atomic Energy Authority, had come into being.

Work on the first power reactor, to be known as Calder Hall, began in 1952, under Christopher Hinton, later Lord Hinton. Construction of the power station commenced in 1953, and only three years later, on 17 October 1956, Her Majesty the Queen opened Reactor 1 at Calder Hall, the world's first industrial scale nuclear power station that could export power to the newly created Central Electricity Generating Board. It had a power-generating capacity of 50 MWe. A second unit was started up in February 1957, and two more in 1958. The reactors operated successfully for more than 47 years and were finally shut down for good in 2003.

The Shippingport power plant, located on the Ohio River in Beaver County, Pennsylvania, about 25 miles from Pittsburgh, was to be the USA's first commercial nuclear power plant based on the pressurised-water reactor. It began operating on 2 December 1957 and was in operation until October 1982. Ground was broken in 1954 at a dedication ceremony attended by President Dwight D Eisenhower, who opened the Shippingport atomic power station on 26 May 1958 as part of the Atoms for Peace Program. The Duquesne Light Company of Pittsburgh built and operated the Shippingport plant. The reactor was a pressurised-water reactor capable of an output of 60 MWe. Specifically, it was a pressurised light-water breeder reactor (PLWBR) designed to run on 93 per cent enriched uranium (normal reactors use no more than 5 per cent enrichment in any fuel rod). The reactor had originally been designed for a large aircraft carrier and was adapted to commercial use.

Meanwhile, the General Electric Company was encouraged by the Atomic Energy Commission to quickly develop a new type of large power reactor that would be cheap and efficient enough to compete successfully with the fossil-fuel-burning electric power plants in widespread use. The complex pressurised-water reactor, with its double cooling loop, was too expensive. As a result GE developed the much simpler boiling-water reactor (BWR). By 1959, the first large boiling-water reactor plant was completed at Dresden, Illinois, and in August of 1960, the first electricity from the 200 MW Dresden generators began to flow into the power grid of the Commonwealth Edison Company, serving the people of Chicago.

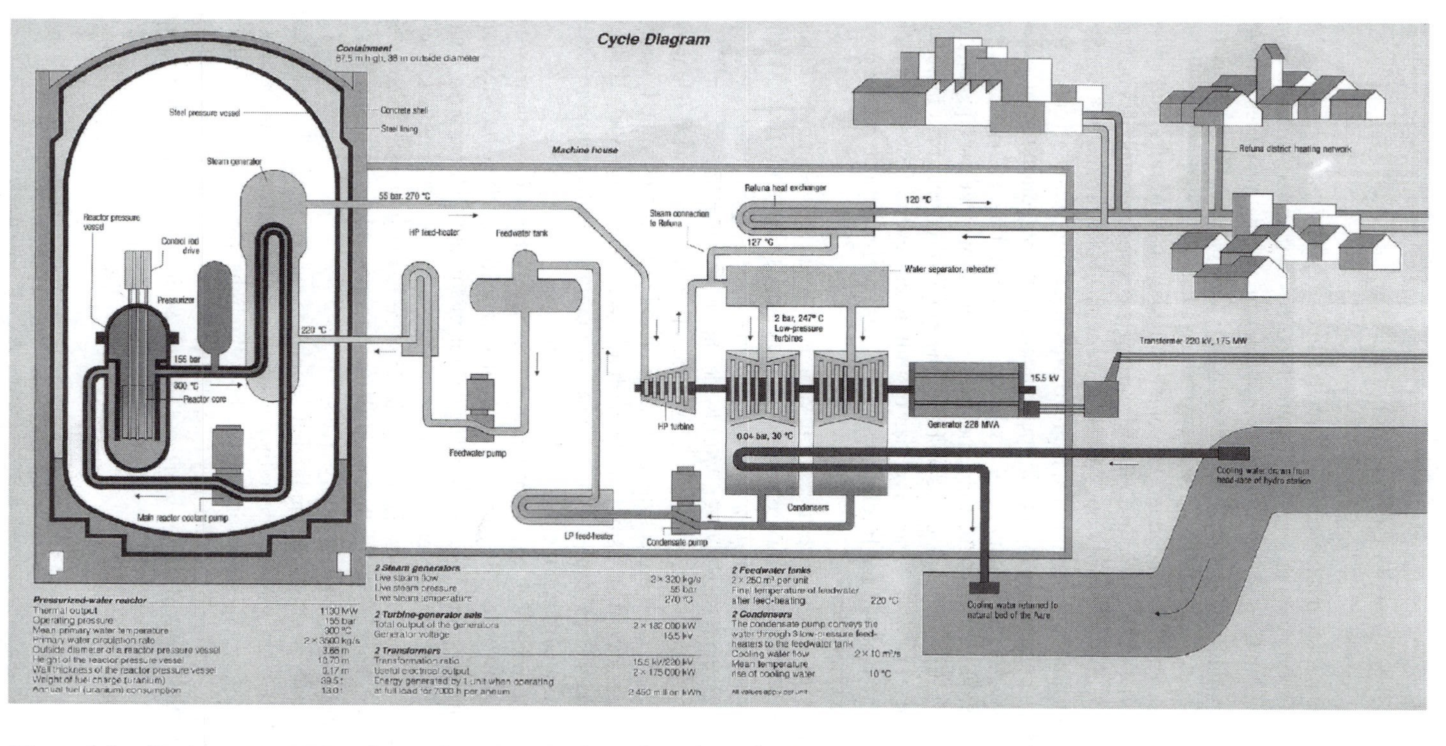

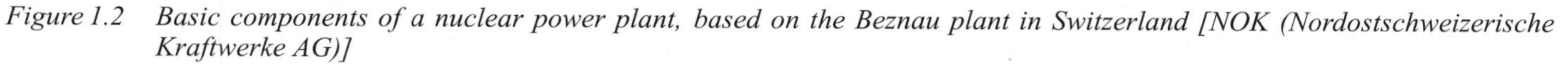

Figure 1.2 *Basic components of a nuclear power plant, based on the Beznau plant in Switzerland [NOK (Nordostschweizerische Kraftwerke AG)]*

Panel 1.3 Sources of radioactivity

Exposure to radiation is inescapable, because there are many natural sources that are encountered every day, both as external sources or internal – inhaled in air or swallowed in food and water. The amount of natural radiation varies depending, for example, on location, but on average it represents about five-sixths of a person's nuclear exposure or around two thousandths of a sievert (abbreviated 2 mSv) per year. The second largest source is medical uses of radiation, which contribute 0.4 mSv per year. Fallout from nuclear testing contributes around 0.02 mSv, while exposure from nuclear power contributes around 0.001 mSv.

Just under half of a person's exposure to natural radiation comes from cosmic rays, which come from sources in space including the sun during solar flares. Cosmic rays affect some parts of the globe more than others: because of the effect of the earth's magnetic field, the poles receive more than the equator, and because the air overhead acts as a shield, the effect of cosmic rays increases with height. The dose at sea level can be multiplied several times for people living 2,000 m above sea level, and air travellers receive a much higher dose, albeit for a short time.

Some of the natural dose arises from various radioactive substances in terrestrial rocks, and this can vary depending on the composition of the rock. Mostly the variation is fairly small, but some 'hot spots' exist, largely associated with granite outcroppings. However, worldwide the most important source of terrestrial natural radiation is radon. This radioactive gas is tasteless, odourless and more than seven times heavier than air and is produced in some rocks by radioactive decay. It seeps out of the earth all over the world but levels in air vary markedly from place to place. In the UK granite rocks like those in Cornwall and Edinburgh are the most important areas.

Radon causes high exposure because it seeps in through the floor of an otherwise airtight building and then becomes trapped – one reason why tropical countries, where houses are designed for maximum airflow, have lower exposure. Since the problem was discovered in the 1970s, when radiation levels more than 5,000 times the typical level were found in houses in Sweden and Finland, housing design has been altered to ensure adequate ventilation to remove radon and reduce activity levels.

Other sources of background radiation include coal burning. Coal's load of radionuclides varies and is generally fairly low, but when it is burnt the radioactive material becomes concentrated in both the light 'fly ash' some of which is carried out through the chimney (the UNEP estimated emissions as 2 Sv per 1,000 MW of coal plant per year) and in the heavier 'bottom ash' left in the furnace.

Of the man-made sources of radiation that contribute to the collective dose, medicine is by far the greatest source and of course it varies from zero to several thousand times the average dose, depending on the patient's treatment needs. Of this, diagnostic x-rays form the greatest proportion, although doses from modern equipment are far lower than those imposed by earlier versions of x-ray machines.

Since 1945 the population has been exposed to radiation from nuclear weapons. Almost none of this arises from the bombs dropped on Hiroshima and Nagasaki. Most arises from atmospheric weapons testing, the two main phases of which were in the 1950s and 1960s.

From the UN Environment Programme *Radiation: doses, effects, risks*

1.4 From research to industry

Within a half century the structure of the atom and its nucleus had been investigated and mapped, and the whole theory of radioactivity and fission had been developed. In the process it had gone from a laboratory curiosity to an industry that could provide power and energy on a commercial scale. In the next 50 years the industry was to see an immense expansion to provide a tenth of the world's power, but also a decline in response to high costs and inefficient operation, and a consistent opposition in some quarters to the industry's existence.

Chapter 2
Reactor designs

2.1 Production reactors

The first large-scale reactors were built for the purpose of producing plutonium for nuclear weapons. Such reactors are generally referred to as 'production' reactors.

The first US production reactor was built at Oak Ridge, Tennessee in 1943, based on the Chicago Pile. A core made of graphite blocks had horizontal channels containing the fuel, which was in the form of cylindrical slugs of natural uranium in an aluminium casing (known as cladding). The channels were cooled by air blown through them.

Eventually nine production reactors were built along the bank of the Columbia River, near the town of Richland, Washington. The Hanford reactors also used natural uranium clad in aluminium stacked in horizontal channels in a graphite core. The heat they generated required high-efficiency cooling, which was achieved by pumping water out of the river directly through the pile and back into the river.

The UK's nuclear programme also began with the impetus to produce nuclear material for military purposes. It began with production reactors, which were built at Windscale in northwest England, on the site of a disused ordnance factory, and later developed into a power programme. The following descriptions of the UK's early reactors are based on Walt Patterson's UK nuclear history *Going Critical*.

As in the US reactors, the first two production reactors used natural uranium clad in aluminium as fuel stacked in horizontal channels in a graphite core, and these were known as Windscale Piles 1 and 2. The cores of the Windscale Piles were made up of a rectangular cross-section stack of graphite blocks pierced by horizontal fuel channels. For fuel loading the channels were arranged in groups of four, access to each group being via a single charge hole in the front shield. In the centre of each group of four channels a single, smaller diameter channel was used for production of other isotopes. Control was via horizontal absorber rods and vertical shutdown rods. Additional vertical channels through the core were provided for experiments in support of the civil nuclear power programme. The uranium metal fuel was in the form of short rods sealed in finned aluminium cans resting in graphite boats. To fuel the reactor, uranium fuel canisters were pushed into channels on one face of the pile, known as the fuel face; as fresh fuel was inserted on one side, it pushed all the canisters along the channel. Following irradiation, fuel rods would be pushed through the rear (discharge) face of the reactor to fall into a water-filled trench from which they would be removed by skips to cooling ponds where, after removal of the graphite boats, they would await processing. The cooling circuit was a once-through forced-air system, the outlet air being discharged through a 137.16 m stack with filters at its top.

The Windscale Piles were cooled by air, blown through the pile at high velocity and discharged through tall stacks. The Windscale Piles were closed following a serious accident in 1954 (see Chapter 4).

Other production reactors went into operation in the then USSR. Five reactors were started up in Chelyabinsk between 1948 and 1957. Five went into operation at Tomsk between 1959 and 1961, and three more at the Krasnoyarsk site in Siberia.

2.2 Graphite reactors

2.2.1 Magnox reactor

The first British power reactors were ’power’ reactors only secondarily. Their other function was to produce plutonium and their operating regime was optimised for this purpose to augment the Windscale output of plutonium. If the purpose of a reactor is to produce fissile plutonium-239, the rate of plutonium production can be ’optimised’ by choice of core geometry, at the cost of other performance characteristics. In addition, the reactor fuel must be changed at relatively short intervals, less than two years on average, otherwise more and more of the plutonium-239 will be converted into plutonium-240, which is not fissile.

With the experience of the Windscale production reactors, the UK chose to follow the same route for its first power reactors, using graphite moderators and carbon dioxide coolant. Four reactors were built at Calder Hall, adjoining Windscale, and four more at Chapelcross in Scotland. The difference was that the heat produced by the fission reaction was extracted from the core and used to generate electricity. Among other changes, this meant that instead of horizontal channels through the graphite, the fuel channels were vertical.

As at Windscale, the fuel was natural uranium, but the ‘cladding’ or skin of the fuel element was an alloy of magnesium, instead of aluminium, which could not be used because its melting point was too low. An additional benefit was that the new alloy absorbed far fewer neutrons than the aluminium, aiding the fission reaction. The new alloy was to be called Magnox, and it gave its name to the two reactors at BNFL (UKAEA at that time) and to the next series of British power reactors. Again for the first time, six fuel elements were held in a single ‘element’ clad in magnox.

In the Calder Hall and Chapelcross reactors, the graphite core, supported on an open framework called a diagrid, was enclosed in a steel pressure vessel, 21 m high and 11 m in diameter, and the vessel itself was enclosed in a concrete shell more than 2 m thick, sometimes referred to as a 'biological shield'. The top of the reactor also had thick concrete shielding (known as the pile cap) but the pile cap also had a number of access points to the fuel channels, for refuelling and servicing.

The carbon dioxide coolant, instead of being discharged, passed through heat exchangers and was then recycled through the core in a sealed circuit. The carbon dioxide coolant was pumped into the reactor from below. The gas passed up each fuel channel, between the fuel element and the moderator. As it passed through the core the gas was heated by the fission reaction, and by the time it reached the top of the channel it had been heated to a temperature of 350 °C. This hot gas was then transferred out of the pressure vessel in pipes that passed through the reactor's concrete shielding and down to four components known as boilers (and also referred to as heat exchangers or steam generators). Here the heat was transferred to a 'secondary' circuit by turning water in the secondary side of the steam generators into steam to turn a conventional turbine generator. The cooled gas was pumped back through the concrete shielding and into the bottom of the pressure vessel to begin its journey through the core again. The gas's circuit through the core was known as the 'primary' circuit. In Calder Hall and Chapelcross the primary circuit had four 'loops', each comprising pump, piping, steam generator and associated equipment. In other designs the number of cooling loops varied.

The carbon dioxide coolant gas was maintained at a pressure of eight atmospheres (117 pounds per square inch or 8.1 bar), so there was special equipment to allow the fuel elements to be changed and other maintenance work to be carried out inside the pressure vessel. The working area above the reactor's pile cap had several moveable 'charging' machines used to insert new fuel and remove spent fuel from the reactor. When spent fuel had to be removed, fresh fuel inserted or when maintenance work was required, these machines were used so that the core did not have to be depressurised.

One machine was used to discharge fuel. It was positioned over an access port, clamped on to the surface of the pile cap and pressurised to match the coolant pressure inside the reactor. Remotely operated equipment inside the machine removed the shielding plug on the pile cap and extended clamps into the core that then lifted out the fuel and stored it in the discharge machine. The shielding plug was then replaced, and the discharge machine was depressurised and removed. The spent fuel is intensely radioactive, containing a mixture of fission products, plus activation products and actinides, along with unfissioned uranium. Still within the discharge machine, it was moved to the plant's cooling pond filled with water, to absorb the radioactivity and cool the fuel for several weeks. In this time the most unstable fission products would have decayed so the radioactivity level drops slightly, and afterwards the spent fuel was transferred to Windscale for reprocessing.

When the spent fuel had been removed from the core, a charge machine, loaded with fresh fuel, would be moved into position and the procedure of clamping to the cap, pressurising and unplugging would be repeated. This time the clamps would lower fresh fuel into position inside the core.

Each sector of the core also had channels for several types of control rod. So-called black control rods containing boron, which have a very high neutron absorbency,

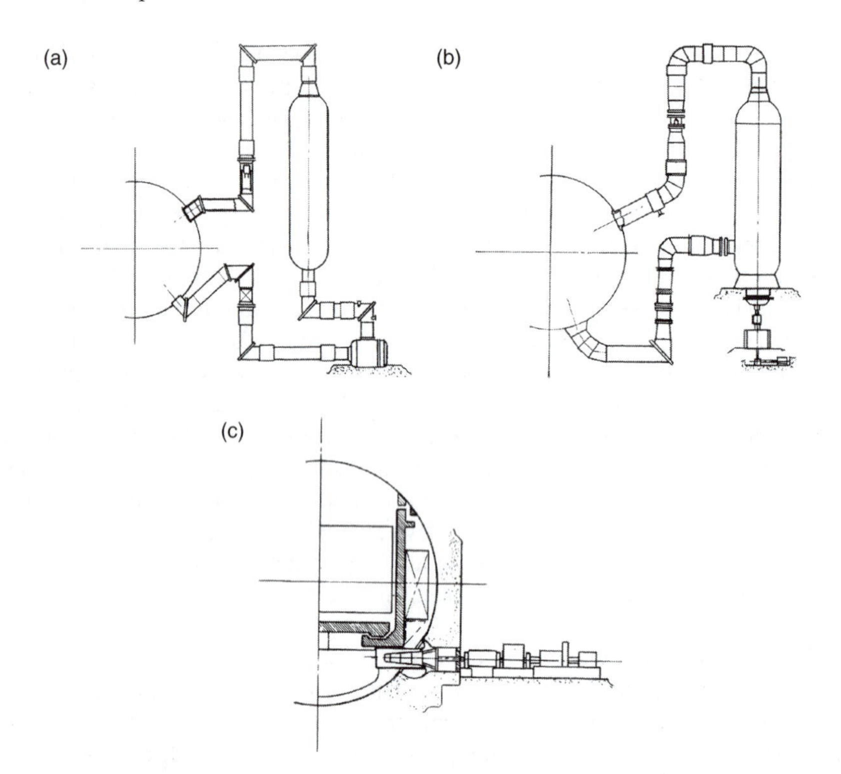

Figure 2.1 Evolution of the cooling circuit in the Magnox series of reactors. Gas circuit (a) Hinkley point, (b) Sizewell, (c) Wylfa [published with permission of Nuclear Engineering International*]*

were used to stop the reaction very quickly and shut down the reactor immediately. These are the 'scram' rods, which were held on electromagnetic grapples, to provide additional 'fail safe'. A continuing electric current was required to hold them above the core: any fault would shut off the magnets and as a result the control rods would immediately fall into the core to halt the fission reaction.

A second set of so-called grey rods acted more slowly and helped control the chain reaction in the core. A third set of control rods were known as 'bulk' rods. These were withdrawn slowly during the 'campaign' or operating cycle to compensate for fuel burnup – the gradual conversion of uranium into fission products that would otherwise gradually 'poison' or slow down the reaction as their concentration in the fuel element increased.

Calder Hall and Chapelcross, similar four-unit plants which were started up across the Scottish border in Dumfriesshire in 1958–9 and finally closed down in 2005, were the design basis for the UK's first generation of commercial nuclear power stations, known as the Magnox series.

The series was announced by the government at the time of concerns over the security of oil supplies following the Suez crisis in 1956. The first stations were to be

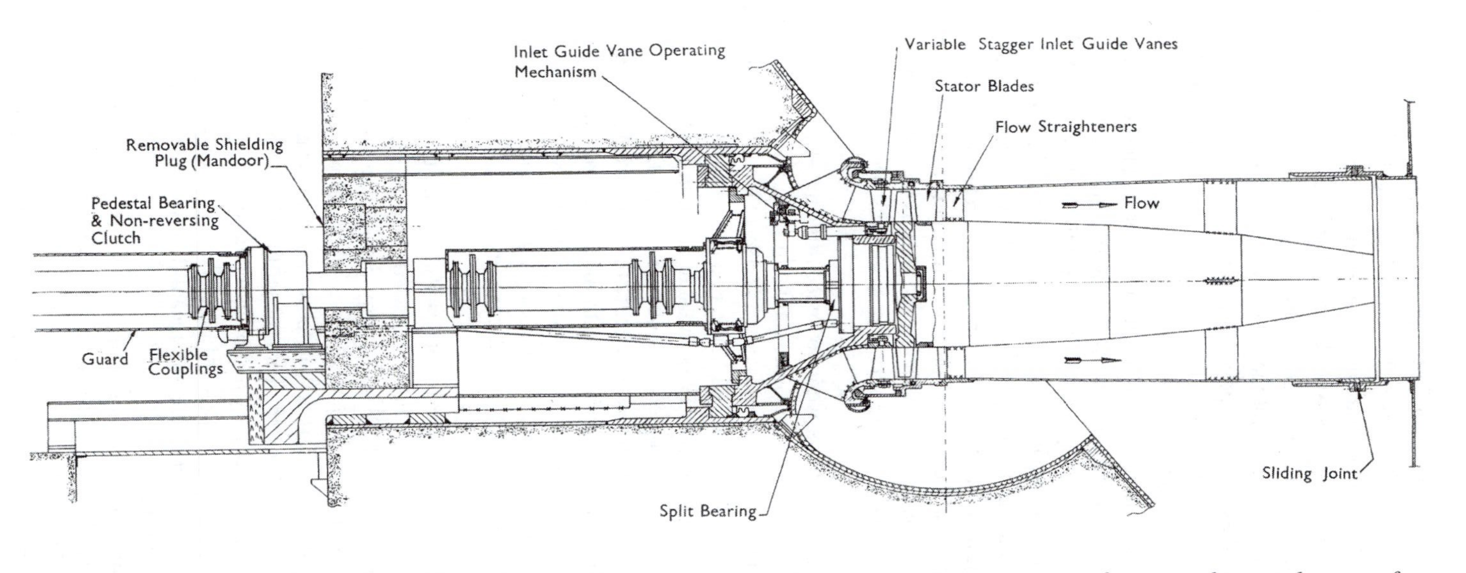

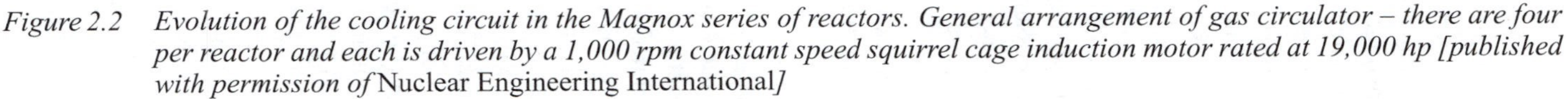

Figure 2.2 *Evolution of the cooling circuit in the Magnox series of reactors. General arrangement of gas circulator – there are four per reactor and each is driven by a 1,000 rpm constant speed squirrel cage induction motor rated at 19,000 hp [published with permission of* Nuclear Engineering International*]*

at Bradwell in Essex and Berkeley in Gloucestershire, where work started in 1957. The two stations reached criticality in 1961 and 1962 and by the end of 1962 all four reactors were in commercial operation and exporting power to the national grid.

In all, eight Magnox stations were built in England and Wales for the Central Electricity Generating Board, and one for the South of Scotland Electricity Board. Each had two identical reactors but the design was developed throughout the series, not least to increase the amount of power produced. The rated power grew from just 129 MWe per reactor at Bradwell and 167 MWe at Berkeley to reach 250 MWe per reactor at Sizewell in Suffolk, which started up in 1966.

The last Magnox station to start up, at Wylfa in Anglesey, more than doubled even Sizewell's power output, producing 500 MWe from each of its twin reactors. The reactor size had grown very quickly in a short time: the core at Wylfa contained nearly 600 t of uranium, compared to just 110 t at Calder Hall. The growing size of the reactor design meant it was too large to use a welded steel pressure vessel employing the technology of the time, so instead pre-stressed concrete pressure vessels were used. A concrete pressure vessel could be made much larger than a steel one, so it made it possible to enclose the steam generators and pumps as well as the reactor's core. At Wylfa the pre-stressed concrete serves both as pressure vessel and as biological shield.

The Magnox power stations were so successful that the design was also used elsewhere. The first to be built overseas, Latina in Italy, was started up in 1964 and operated for 23 years until it was shut down in 1987. Japan's sole Magnox station, Tokai, was rather smaller, rated at 166 MWe, and it was a joint construction between GEC and Fuji. It was started up in 1965 and shut down in 1998.

France had also initially adopted gas-cooled reactors. By 1971 it had seven gas–graphite reactors in operation, the largest being the St Laurent A2 reactor at Orleans, which generated 465 MWe. France was instrumental in providing a Magnox design to be built at Vandellos in southern Spain, which was started up in 1972 and operated until 1990 when it was closed following a fire in the turbine hall (separate from the nuclear part of the power station). It had been one of the largest Magnox stations, rated at 496 MWe.

The rate of conversion of heat to electricity and the power generated by each kilogramme of uranium fuel had both increased through the Magnox series in the UK. Calder Hall produced 2.4 kW of energy per kilogramme of fuel and its 268 MWt heat output was converted to 50 MWe of electrical power. But at Wylfa, each kilogramme of fuel produces 3.16 kW of power, and its heat output of 1,875 MWt is converted to 580 MWe of power. That meant the efficiency of conversion of the heat from the reactor into electricity had been increased by more than 10 per cent to over 30 per cent. Nevertheless the low rate of conversion meant the reactors were relatively inefficient – and therefore that meant they were expensive to operate and required more uranium fuel.

One way of improving the energy efficiency of the reactor was to transfer more energy from the core by increasing the coolant temperature. But in the Magnox design, the Magnox fuel cladding determined the maximum temperature allowed in the core. If the coolant temperature was too high it could melt the cladding and could also distort the uranium fuel. Incremental improvements would not overcome this barrier, and a new design was needed if a more efficient reactor was to be built.

2.2.2 Advanced gas reactor

The second generation of British power reactors aimed for much better heat and power outputs. As a result they used a different fuel: uranium dioxide, with stainless steel cladding. These materials have a much higher melting point than the Magnox design. However, a drawback in the use of stainless steel cladding is that it tends to absorb neutrons much more effectively than the magnox cladding, slowing down the chain reaction. To offset this effect, the uranium used in this new fuel design is slightly enriched. Therefore, in the second generation so-called Advanced Gas Reactors (AGRs) the proportion of uranium-235 increased to about 2 per cent. As with the Magnox reactors, the moderator in the AGRs is graphite.

The design of the fuel element was also changed to improve heat transfer. The stainless steel tubes containing the uranium dioxide fuel are just a couple of centimetres in diameter and a metre long – very similar to those used in a PWR (see below). A set of tubes containing the fuel are grouped in a 'fuel assembly' (sometimes referred to as a fuel rod) with spacer grids at intervals to ensure they are kept separate in operation. The assembly is held in a graphite cylinder, through which the coolant passes. The second benefit of the change in fuel design was that it made the AGR relatively compact, when compared to the Magnox core.

The philosophy behind the AGRs was to greatly increase the heat transfer from the fission reaction by using a much higher coolant temperature, to improve thermal to electric conversion efficiency. As in the final Magnox design at Wylfa, the AGR design once again uses a pre-stressed concrete pressure vessel that encloses core, steam generators and pumps. As with the Magnox series, access is from above, through the pile cap, and control rods therefore enter the core from above. Refuelling is done using a single machine, which performs both the loading of fresh fuel and the unloading of spent fuel. The fuel elements, stacked eight to a vertical channel, are coupled together, removed and inserted as a single long string; accordingly the refuelling machine is very tall – equivalent to a four storey building.

The AGR was designed to have a thermal efficiency of around 40 per cent, as it finally achieved very high coolant temperatures of well over 600 °C.

The UK Atomic Energy Authority (AEA) won government go-ahead to develop its first AGR in 1957. It was to have a heat output of 100 MWt, to produce 28 MWe of electricity (it was later found to be more efficient than expected and its capacity was increased to provide 36 MWe of electricity). The eventual plant was sited at Windscale and was called the Windscale Advanced Gas Reactor, and known as WAGR. Construction was completed in January 1963 and in the following month it was already supplying power to the national grid – ten years before the last Magnox station went into operation.

The full-scale AGR construction programme was started in 1965. The first to be ordered was a twin-unit station, Dungeness B, but this was also the most problematic in construction, and by the time it was operating commercially, twenty years later

scale: 1 in. = 68 ft.

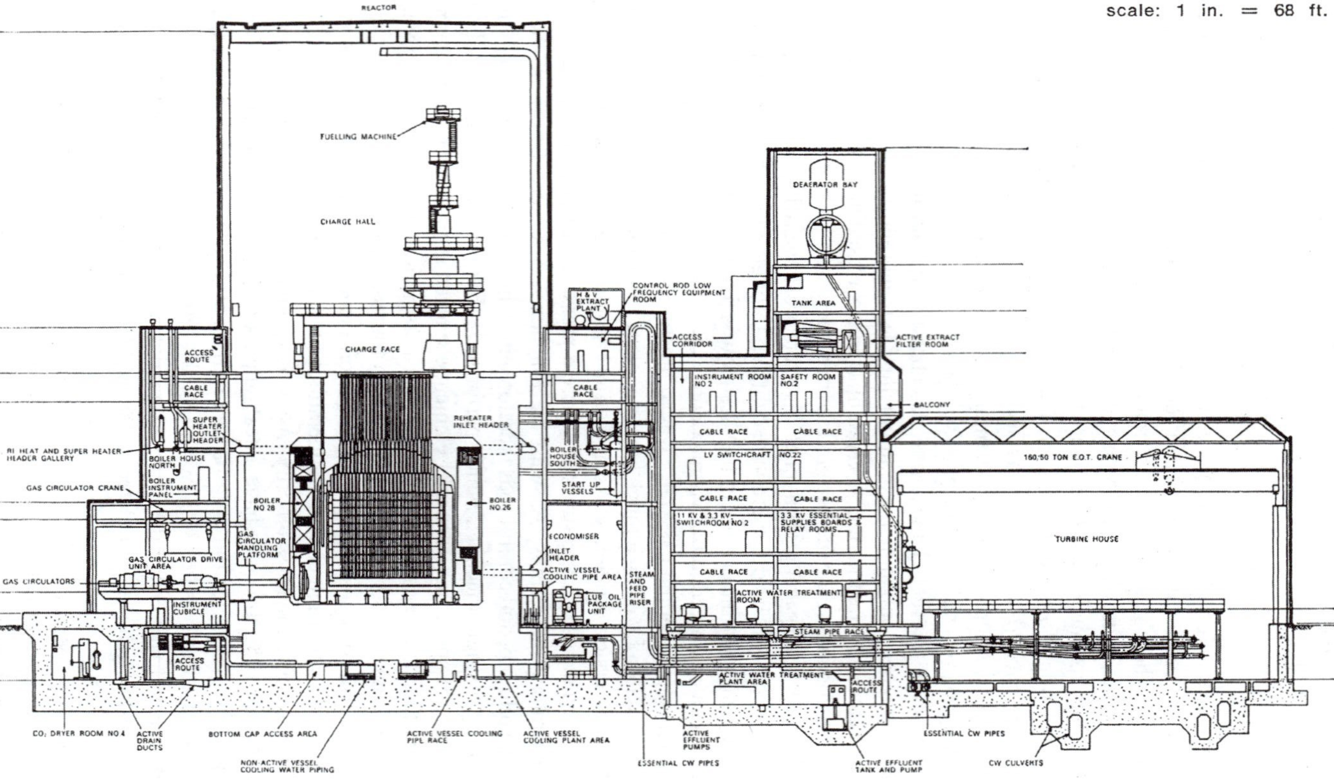

Figure 2.3 Dungeness advanced gas reactor (AGR) [published with permission of Nuclear Engineering International*]*

in 1985, it had been overtaken by almost all the other AGR plants. These were Hartlepool, Heysham 1, Hinkley Point B and Hunterston B (all of which had two reactors). The final two AGR stations, at Torness and Heysham 2, in contrast, began construction in 1980 and took nine years to reach commercial operation.

The UK took the AGR as its chosen reactor style, but never standardised the design as the reactors were developed by different consortia using design data from the UK AEA. As a result, the AGRs never benefited from the expertise and experience that could be developed in a standard design series.

2.2.3 RBMK (high-power channel reactor)

The USSR also developed graphite-moderated reactors, known as RBMKs (abbreviated from the Russian Reaktor Bolshoi Moshchnosty Kanalny, or 'high-power channel reactor'), developed from its production reactors.

The RBMK reactor has a cylindrical graphite core up to 7 m high and 12 m in diameter. The core contains around 1,700 vertical pressure tubes that contain fuel, or control or shut down assemblies. The fuel is uranium oxide, enriched to 2 per cent uranium-235.

Unlike the Magnox or AGR, the cooling medium in the RBMK is water, which is passed through the reactor via the 1,700 pressure tubes. The water is allowed to boil as it passes through the core, as with the boiling-water reactor (see below), and because the tubes are at high pressure boiling occurs at 290 °C.

The reactor has up to eight coolant loops. In the coolant loops water is pumped to large headers and from there to smaller headers that serve groups of feeder pipes. The coolant flows up the fuel channels and is allowed to boil. Once it has left the core it passes through a steam separator: the steam drives the power turbine, while the water in the mixture (along with water on the return leg from the turbine) is fed back into the circuit.

The reactor core is inside a steel vessel, which is filled with a mixture of helium and nitrogen to maintain an inert atmosphere around the graphite core (which can be at a temperature of around 600 °C). The reactor is controlled with around 200 boron carbide absorber rods, of which around 24 are emergency protection or 'scram' rods. In addition, there are a variety of different 'local automatic regulation' rods, power regulation rods and 'bottom entry' rods that are used to maintain the power profile across the large core.

The reactor and most of the coolant circuits are inside thick-walled leak-tight enclosures linked to pressure suppression chambers. The top of the reactor is covered with a 3 m thick shielding slab made of steel and concrete. The slab has access points to allow access to the fuelling channels and, as at the UK's gas reactors, refuelling is performed on-load using a refuelling machine that moves on the top of the shielding.

After completing two prototype reactors at Beloyarsk, Ekaterinburg, the USSR built 11 RBMK reactors in Russia. They are sited at Kursk, St Petersburg and Smolensk (four small reactors based on a variant of the technology at Bilibino). Each reactor produced some 1,000 MWe of electricity and four similar reactors were built at Chernobyl in Ukraine. Two more reactors of a still larger design, rated at 1,500 MWe for each unit, were built at Ignalina in Lithuania. Plans for two additional RBMKs at Kursk and Smolensk were cancelled after the Chernobyl accident (see Chapter 4), as was a third 1,500 MWe unit at Ignalina. In addition, work was halted on two further reactors that were already under construction at the Chernobyl site.

Since the accident in 1986, the four reactors at Chernobyl have been shut down, but they were in operation for several years after the accident. Even unit 3, which shared a turbine hall and common wall with the destroyed Chernobyl 4, operated for several years after the accident in 1986. Chernobyl 2 operated until there was major damage to the non-nuclear turbine hall in a fire in 1992, while Chernobyl 1, the last operating reactor on site, was shut down in 1996, ten years after the accident at the fourth unit.

2.2.4 High-temperature gas reactor

The movement from Magnox to AGRs in the UK aimed to abstract much greater power from an equivalent amount of uranium, by increasing the heat transferred. It was clear that operating reactors at even higher temperatures (about 950 °C) would permit not only higher efficiency, but could also lead to wider direct applications such as hydrogen production, which is more efficient at higher temperatures. The continuing attempt to increase operating temperatures led to the high-temperature gas reactor, or HTGR.

HTGRs have been built and operated as test reactors and are the basis for several new designs currently being developed.

The very high achievable coolant temperatures lead to high steam cycle efficiencies, or alternatively, make possible the ultimate use of gas turbines directly driven by the coolant.

The first high-temperature gas-cooled reactor in operation was the Dragon reactor at the UK Atomic Energy Authority reactor development site at Winfrith, in Dorset.

HTGR fuel is not contained in fuel pins clad in metal as in LWR and LMFBR fuel, but in fuel particles. The fuel is in the form of tiny spheres (less than a millimetre in diameter). Each has a kernel of uranium oxy-carbide, with the uranium enriched up to 9 per cent ^{235}U, surrounded by layers of carbon and silicon carbide, giving a containment for fission products which is stable up to 2,000 °C. In order to retain fission products, each particle is enclosed by several coatings of ceramic material with a high-temperature stability. The particles are homogeneously dispersed in a graphite matrix that is subsequently compressed to spherical elements, pebbles or in the form of rods, filled into fuel channels of a multi-hole graphite block. The particles remain intact and retain virtually all fission products up to a temperature of about 1,600 °C. They do not melt at a given threshold temperature and fail only gradually under accident conditions; hence, a sudden release of fission products cannot occur.

Steam and carbon dioxide may react with graphite in such conditions, so helium is generally chosen as the coolant.

A cylindrical core contains the fuel elements and is surrounded by a moderating graphite reflector. The control rods are located vertically in the circumferential wall.

Helium gas is heated by passing it over the fuel in the core, which in more advanced designs drives a gas turbine in a direct cycle to produce electricity.

Further HTGR development will concentrate on improved plant performance and life extension studies. With respect to the former, much effort is currently devoted to the so-called gas-turbine-cycle in which high-temperature gas is conveyed directly to a gas turbine, yielding very high thermal efficiency and promising low energy cost, and to production of very high-temperature process heat.

The high-temperature gas reactor is the basis of several designs now under development. The pebble-bed reactor being developed by a consortium is expected to be ready for deployment in the next few years. In the long term, high-temperature reactors may be used to produce hydrogen as well as power, and such designs form an important part of the current US research effort.

2.3 Light-water reactors

While European designers in the UK, France and Russia were pursuing reactors that used graphite moderators, the US was leading development of reactors that used 'light'-water moderators.

Light water is in fact ordinary water (H_2O), but is usually referred to as such to distinguish it from so-called 'heavy' water (D_2O), where hydrogen is replaced by deuterium, and which is elsewhere used as a moderator, for example in CANDU reactors (see below).

Light-water reactors have been the world's most widely accepted power reactors. In 2005 various light-water reactors accounted for 367 of the 447 reactors in operation worldwide, 17 of the 28 under construction and 30 of 38 at various stages of planning.

2.3.1 Pressurised-water reactor

The pressurised-water reactor (PWR) got its start in the development of propulsion reactors for US nuclear submarines. In addition to its extremely infrequent need to refuel (less than annually), high power and long range, 'burning' nuclear fuel did not require oxygen, unlike diesel engines, so nuclear-powered submarines could remain submerged indefinitely. The PWR design, therefore, was dictated by the strict space limitations inside a submarine.

US designers developed a reactor that used a small core with high power density, immersed in a tank of light water, which is used both as coolant and moderator. The fuel is in the form of clusters (assemblies) of enriched UO_2 rods, which may be of the order of 1–2 cm in diameter and up to 7 m in length. The uranium dioxide fuel is clad in zirconium alloy or, in some cases, austenitic stainless steel. The fuel assemblies are square in shape, i.e., the rods form a square array in each fuel assembly, with the assemblies, in turn, being closely packed in a square array forming the reactor core. Each array may have up to 280 fuel pins in it (in a 17×17 array), which are separated by spacer grids to allow the coolant water to pass between the fuel pins.

The core is located in a relatively small steel pressure vessel that may in a 1,000 MWe power reactor be typically 3–4 m in diameter and 12–13 m high. The pressure vessel is made of cast steel and is around 20 cm thick. The steel is further clad with a layer of stainless steel typically 5 mm thick to stop it from being corroded. Control rods and their drive mechanisms are included within the reactor pressure vessel, in the space above the core, and their controls are fed through the vessel lid or 'head'.

The water coolant at high pressure (around 2,000 psi) is circulated by external pumps into the reactor vessel and flows upwards through the fuel clusters. It is pumped out of the vessel through large pipes in the side of the pressure vessel above the level of the core, to steam generators and from there back to the pumps. The cooling loops, including the recirculation pumps and steam generators, are the reactor's primary circuit, in which water is recirculated through the core. Water in the primary circuit does not boil because it is at too high a pressure. Instead, it gives up its heat to water in the secondary circuit via the steam generators.

On the secondary side of the heat exchangers the water passing through is at a lower pressure and thus it boils, forming saturated steam, which drives the turbine. This steam is generated at around 750 psi, leading to a relatively low overall station efficiency (around 30 per cent).

In order to refuel the reactor, it must be shut down, cooled and depressurised. The top of the pressure vessel is then unbolted and removed and the fuel assemblies changed. An entire or part core may be replaced and each fuel assembly may remain in the reactor for up to two years, depending on the operating regime.

In order to operate for long periods without refuelling, the new fuel is relatively highly enriched in ^{235}U, and may be enriched by anything from 1.7 to 4.4 per cent, depending on the operating regime. Such high enrichment does mean there is greater reactivity in the core in the early stages before so-called poisons (unfissionable daughter products from the reaction) begin to arise in the core. While the fuel is new, the excess reactivity in the core may be compensated for by a neutron poison dissolved in the coolant/moderator water. Alternatively, as fuel residency in the core tends to increase, the enrichment reload fuels may now be designed with a small proportion of 'burnable poisons'. These are isotopes that initially slow down the reaction but are gradually converted to isotopes that have a neutral or positive effect on reactivity as time passes.

PWRs come in several variants with different numbers of cooling loops: Westinghouse, for example, has produced two-, three- and four-loop designs. Combustion Engineering produced a two-loop design and Babcock and Wilcox's design had five loops.

Part of the UK's initial opposition to the PWR design was that active pumping was required to maintain water flow over the core, and hence cooling. A so-called loss of coolant accident could uncover the core and allow the heat of the core to increase so far that it would melt. UK engineers were also concerned that welds required to manufacture the reactor pressure vessel and in the primary circuit would be weak spots in the circuit. This was still being raised as a concern when the Sizewell design was being examined in the 1980s.

However, the reactor pressure vessel and primary circuit (including the steam generators) are surrounded by a so-called containment – a leak-tight steel shell surrounded by thick concrete shielding. The containment is intended to trap all the core contents and fission products in the event of an accident, up to and including a pipe break in the primary circuit, and loss of coolant accident, that would leave the core uncovered. The containment was tested at Three Mile Island where a core meltdown was retained inside the containment with no radiological release detected off the plant site (see Chapter 4).

2.3.2 The USSR's PWR – the VVER

The USSR was as interested in developing reactors for its submarines as was the USA, and for very much the same reasons. As a result, alongside its RBMK programme the USSR developed a pressurised-water reactor referred to as a VVER.

The basic principle of the VVER was identical to that of the PWR. A compact core was made up of uranium dioxide fuel assemblies, comprising fuel rods not much more than 1 cm in diameter and more than 2 m long. In contrast to the square array used in the PWR (and BWR) fuel in the VVER fuel assembly was held in a hexagonal grid.

As with the PWR, light water used as both coolant and moderator was passed at very high pressure through the core, in a small reactor pressure vessel around 12 m high and 4 m in diameter.

Information from the VVER at Paks in Hungary describes the design and how different it is from the Western PWR. The fuel of the VVER reactor is uranium dioxide (UO_2), which is compacted to cylindrical pellets of about 9 mm height and 7.6 mm diameter. In the centreline of the pellets there is an inner cylindrical hole of 1.6 mm. This allows gaseous fission products to escape, reducing the pressure, but it also reduces the fuel temperature. The uranium pellets are inserted into a 2.5 m long and 9 mm diameter tube made of a zirconium alloy, which is sealed hermetically. The cladding prevents fission products and other radioactive material from getting into the cooling water.

The VVER-440 fuel assembly is a hexagon and each contains 126 fuel rods. The fuel enrichment can be 1.6, 2.4 or 3.6 per cent, but normally all the rods in an assembly are of the same enrichment. Altogether 349 assemblies can be inserted into the reactor core and 312 out of these are fuel assemblies.

The reaction is controlled with absorber rods made of borated steel. The control rods are inserted into the reactor from the top. There are altogether 37 such control rods, out of which 30 are always pulled out of the core during operation. These are the safety rods, which provide emergency or fast shutdown. The other seven absorbers are used to control power during operation.

The reactor fuel is rotated over time for maximum efficiency. In a refuelling outage a year after it starts up those used fuel assemblies that in the first core had contained 1.6 per cent enriched fuel are removed and 2.4 per cent enriched assemblies are put in their place. The originally 3.6 per cent enriched assemblies are put into the place of the 2.4 per cent ones and their place is filled with fresh (3.6 per cent enriched) fuel assemblies. When the power plant is shut down for its annual refuelling outage, the spent fuel is removed. The other assemblies are moved according to the above description, while fresh fuel is put into the outer places. So with the exception of the initial loading, each assembly spends three years in the reactor.

The reactor core is held in the reactor pressure vessel, which is a vertical cylinder of total height 13.75 m and outer diameter 3.84 m. The vessel is made of steel; its thickness at the height of the reactor core is 14 cm and there is an inner 9 mm-thick stainless steel plating (clad lining) as corrosion prevention. There are six inlet and six outlet pipe connections located at different heights on the vessel.

The heat generated in the reactor core is transferred via six cooling loops, which surround the reactor. In each cooling loop, water heated to 297 °C exits the reactor and is piped to the steam generator. At 2.3 m diameter and 12 m length the steam generator is similar in scale to that of a PWR, but its configuration is different: the cylinder in a VVER is horizontal, whereas in a PWR it is vertical.

In the steam generator the radioactive primary circuit water flows through 5,536 heating pipes, each of 16 mm diameter, resulting in the boiling of the secondary side water. The cooled water returns to the reactor in the primary circuit. Water in the primary side is circulated by the main circulating (coolant) pump. Each coolant loop can be separately closed using a valve. It is the pressuriser's (or expansion tank's) task to keep the pressure at a constant value of 123 bar so the water in the primary circuit does not boil. Each unit has one pressuriser, which is connected to the hot leg of a loop. If the primary pressure decreases, water might start to boil. In order to prevent this, electric heaters switch on automatically in the pressuriser. Due to the heating there will be intense boiling, more steam will be generated and this leads to a pressure increase.

The primary water, which circulates in small tubes at a temperature of 297 °C, causes the feed-water that enters the steam generator at 46 bars and 223 °C to boil. The moisture content of the generated steam of 260 °C must be lowered, otherwise the turbine blades would be damaged. This purpose is served by moisture separator shutters put into the way of steam. When steam passes through these plates, water drops deposit and thus the moisture content of the outgoing steam will only be 0.25 per cent.

The steam leaves the steam generator at a mass flow rate of 450 t/h and heads towards the turbine. Out of the six steam generators three feed a turbine in a given unit (there are two turbines for each unit).

In case of a failure, all control rods are automatically inserted into the reactor core and within a few seconds they stop the chain reaction. However, a significant amount of heat is produced due to the decay of radioactive fission products and this heat production can be as high as 7.5 per cent of the nominal power in the first moments. Therefore, cooling of the core is absolutely necessary. If the cooling system is damaged emergency cooling must be provided, even after the reactor is shut down.

The most serious design failure of the nuclear power plant is the rupture of a primary circuit main pipe. In the case of such a serious accident the loss of normal core cooling is worsened when the water at the rupture, which is at a very high temperature with low outside pressure, immediately starts to boil and floods the surrounding space with highly radioactive steam.

In a pressurised-water reactor such an accident is retained inside the reactor building by the containment. The VVER uses an alternative so-called hermetic room and localisation system.

The hermetic room is a part of the building, with 1.5 m thick concrete walls, which on the one hand shields against radiation and, on the other hand, prevents steam from getting out up to 1.5 bars pressure. In order to avoid higher steam pressures, a steam pressure reduction system has been developed, which consists of a localisation tower and a sprinkler system. Steam produced along with the air of the hermetic room flows to the localisation tower, where it flows through trays filled with water. Meanwhile, steam condenses and thus the pressure of the hermetic room drops. The sprinkler system sprays borated water into the hermetic room. This causes further pressure decrease.

Boric acid is necessary because the condensed water may later get into the reactor and the neutron absorbing ability of boron helps to prevent the chain reaction from starting again.

The lifetime of the reactor is determined by the lifetime of the pressure vessel. However, the configuration of the VVER-440 means that welds in the vessel are subject to damage from the constant neutron radiation, which gradually changes the structure of the weld. The problem is unique to VVER-440s because of the weld position. The damage can be reversed by a heat treatment on the weld known as annealing, and the effect can be lessened by altering the management of fuel assemblies and using older assemblies in the outer edges of the core.

There are two main generations of VVER-440 (so called because it produces 440 MWe of electricity), denoted V-230 and V-213, and some other versions employed in Russia itself.

Russia has six VVER-440s of various generations in operation and supplied a number of them to countries in Eastern Europe, then members of the Comecon block. Four are in operation in Bulgaria, four in the Czech Republic, four in Hungary, six in Slovakia and two in Ukraine. There are also two (albeit with many design changes) in Finland and one in Armenia (where a second unit has been shut down).

The USSR began work on a much larger VVER – the VVER-1000 – in the late 1960s and began to deploy it in the early 1970s. This reactor is also known as the V-320 and it is much more similar to western-style PWRs than the smaller VVER-440. It retains the configuration of horizontal steam generators (western-style PWRs have two to four vertical steam generators), which its Russian designers claim are much less prone to ageing than the Western vertical units, and so far this has been borne out by experience. However it uses four steam generators, instead of the smaller version's six.

One major change in the scale-up to the VVER-1000 is that its Russian designers abandoned the tower or 'hermetic room' pressure relief system, in favour of a single sealed containment similar to those used in the PWR.

There are nine VVER-1000 reactors in operation in Russia, at Nonovoronesh, Kalinin and Balakovo.

Two VVER-1000s are also in operation at Temelin in the Czech Republic, and 13 in Ukraine at Khmelnitski, Rovno, Nikolaiev and Zaporozhe. Most of the

VVER-1000s started up in the 1980s and plans for more units were cancelled following the break-up of the Soviet Union. The effects of the Soviet breakdown also halted work on several units that were at various stages of completion – some, notably Rovno 4 and Khmelnitski 2, both of which are in the Ukraine, were almost ready to start up. Those two reactors have now been completed and are in operation. So has the two-unit station at Temelin in the Czech Republic, although as there is extensive nuclear expertise and fabrication capacity in the Czech Republic that plant was less affected by the USSR's collapse, and delays were mostly due to a variety of other technical and political factors.

In Bulgaria, where the four VVER-440s in operation at Kozloduy are among the oldest in the world and are due for closure as Bulgaria prepares to join the European Union, the government had long planned two VVER-1000s at a site on the Danube called Belene. Civil construction work was well underway but work has been more or less suspended since the early 1990s despite intermittent government support.

2.3.3 Boiling-water reactor

The BWR (boiling-water reactor) is second only to the PWR in terms of worldwide acceptance. It is similar in many respects to the PWR, the basic difference being that the light-water coolant is allowed to boil in the reactor core.

It was known at the start of reactor design that if the water in the core was allowed to boil it would be more effective in removing the heat of the fission reaction and converting it to power. However, at that time boiling was thought likely to trigger instabilities in a reactor core. A phenomenon called nucleate boiling was allowed in the PWR – very tiny bubbles of gas – but that was all.

It is true that a void or gas bubble caused by boiling water turning into steam can affect the reaction in the core; however, this can be varied according to the configuration and operation of the reactor core and reactors can be designed so as to have a positive or negative void coefficient. In the former case a void speeds up the reaction and in the latter it slows it down. A negative void coefficient is generally preferred in reactor design, for safety reasons, but in some cases where fuel containing poison is used, a positive void coefficient is employed in the early stages of the fuel residency in the core, until its reactivity increases.

Research in the 1950s soon led to an early BWR design, produced by the Argonne National Laboratory. A prototype BWR, Vallecitos, ran from 1957 to 1963. The first commercial boiling-water reactor was Dresden 1, which was rated at 250 MWe of electrical power and which was designed by General Electric. The reactor was started up in 1960. The first US town powered by nuclear energy – Arco, Idaho, population 1,000 – was powered by an experimental version of the boiling-water reactor known as Borax III.

BWRs were originally designed by Allis-Chambers and General Electric (GE). The General Electric design has survived, whereas all Allis-Chambers units are now shut down. The first US commercial plant produced by General Electric was at Humboldt Bay (near Eureka) in California. Other suppliers of the BWR design worldwide included ASEA-Atom, Kraftwerk Union and Hitachi. Commercial BWR reactors

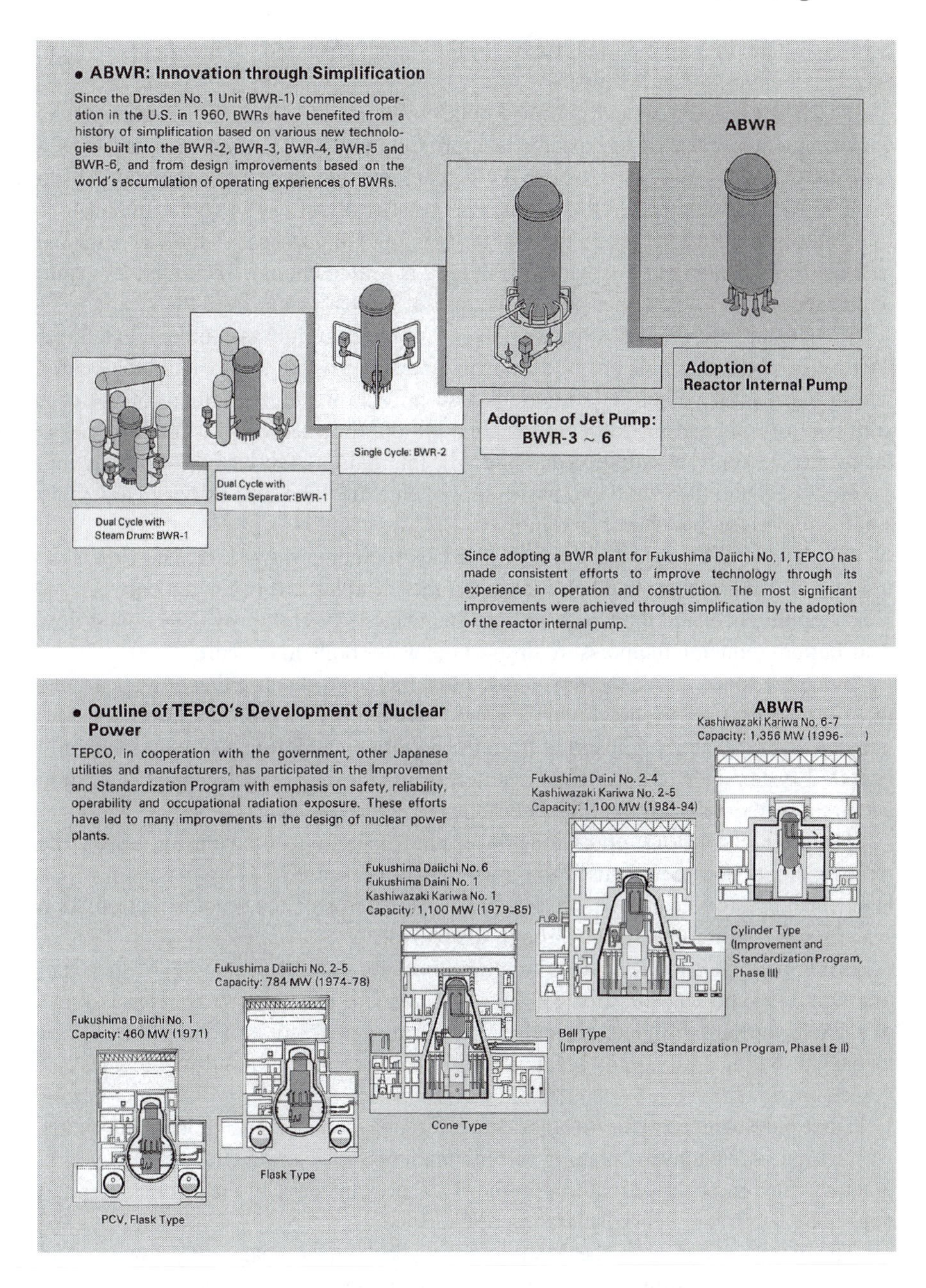

Figure 2.4 Evolution of the boiling-water reactor as deployed by Tokyo Electric Power Company [published with permission of TEPCO from their brochure Advanced Boiling Water Reactor*]*

may be found in Finland, Germany, India, Japan, Mexico, Netherlands, Spain, Sweden, Switzerland and Taiwan.

The BWR reactor typically allows bulk boiling of the water in the reactor. The operating temperature of the reactor is approximately 570 °F, producing steam at a pressure of about 1,000 psi. Current BWR reactors have electrical outputs of 570 to 1,300 MWe. As with the PWR designs, the units are about 33 per cent efficient.

Water is circulated through the reactor core, picking up heat as the water moves past the fuel assemblies. The water eventually is heated enough to convert to steam. Steam separators in the upper part of the reactor remove water from the steam.

As with a PWR, the BWR has a small core made up of fuel assemblies. In a BWR the reactor vessel internals provide the structural support for the core and fuel located within the reactor vessel. The internals also provide for the alignment of the core with control rods and instrumentation. In addition, they serve as a shield to protect the reactor vessel from radiation damage. The internals are fabricated from austenitic stainless steel and they have two basic components: the core support barrel assembly and the upper guide structure assembly.

In addition to the two assemblies, internals include a flow skirt. The flow skirt is a cylindrical structure that has a large number of holes and is located between the core support barrel and the lower head of the reactor vessel. It provides coolant flow distribution, reducing the pressure drop as the water boils in the core.

Since the space above the BWR core must include a steam-collecting apparatus, the control rods, used to shut down the reactor and maintain a uniform power distribution across the reactor, are inserted from the bottom by a high-pressure hydraulically operated system. The control rods inserted from below are cross-shaped in section and are inserted between the fuel assemblies.

The BWR also has a suppression pool around the bottom of the reactor vessel. The torus or suppression pool is used to remove heat released if an event occurs in which large quantities of steam are released from the reactor or the reactor recirculation system, used to circulate water through the reactor.

BWR fuel, like PWR fuel, is uranium dioxide, enriched to some 2.5 per cent uranium-235, and clad in zircaloy. The fuel pins are not quite so slender as those of a PWR, and are grouped, typically, in an array open-ended to allow the coolant to pass through, and supported above and below on plates fixed to the pressure vessel.

Since the steam fed to the turbines comes directly from the BWR core, it may carry with it some radioactivity, making turbine maintenance more difficult. However, in practice, most radioactive material in the BWR coolant stays in the liquid water and does not get carried over by steam into the turbine.

The BWR shares with the PWR the drawback of comparatively low coolant temperature and resulting inefficiency of conversion of heat to electricity. It also shares the problems associated with relatively high power density, although the power density of a BWR is likely to be only half that of a comparable PWR. The BWR is also potentially more susceptible than the PWR to the possibility of 'burn-out' or 'steam blanketing', which arises if a layer of steam is allowed to form next to the hot fuel. The low conductivity of the steam would mean that the heat would no longer be so effectively removed from the fuel, and the fuel temperature might then rise suddenly and dangerously.

The steam thus produced is separated from the coolant water by centrifugal separators located in the reactor vessel above the core and fed directly to the turbine at ~1,000 psi pressure.

BWRs have been almost as successful as PWRs in being deployed worldwide. There are 35 units in operation in the US. In Europe, there are six in Germany, and eight in Sweden, and they are or were also used in Finland, Italy and Spain. The BWR has also been successful in Japan, where 28 units are now in operation and more are planned.

2.4 Heavy-water reactors

The Canadian role in wartime fission research had been particularly concerned with heavy water. But after the war, Canada decided against a nuclear weapons programme. Accordingly, with no facilities for enrichment but with substantial supplies of uranium, Canada chose to concentrate on heavy-water natural-uranium reactors.

The result was the development of a Canadian power reactor known as the CANDU reactor (from CANadian Deuterium Uranium).

2.4.1 Canadian deuterium uranium reactor

The basis of the CANDU reactor is a horizontal cylindrical tank with hundreds of horizontal tubes running completely through it from one side to the other. This tank, called a 'calandria', is filled with heavy water at atmospheric pressure, kept well below boiling point by its own cooling circuits. The barrel and ends of the calandria are of stainless steel; the tubes are of zirconium alloy. Running through these calandria tubes there are zirconium alloy pressure tubes that contain natural uranium dioxide fuel rods in short clusters (50 cm long). The pressure tubes are each connected to inlet and outlet feeders. The coolant, which in most designs is heavy water, passes through the zirconium pressure tubes. The space between each pressure tube and the calandria tube that encloses it is filled with carbon dioxide to insulate the calandria, which is full of heavy-water moderator, from the hot coolant tubes.

The heavy-water coolant at some 90 atmospheres pressure passes through heat exchangers and pumps and generates steam to drive turbines; the system beyond this point is very similar to that in a light-water reactor.

The reactor power is varied by varying the level of moderator in the calandria; less moderator means fewer fissions. The reactor can be scrammed by dumping all the moderator through large valves into a tank below the reactor; the valves are normally held shut by helium pressure, and any reactor fault will de-energise them so they open immediately in a 'fail safe' way.

Refuelling is done on load, with the reactor in operation, by means of machines which couple on to either end of a pressure tube. One machine pushes fuel in at one end of the tube, which pushes the last bundle of irradiated fuel at the other end of the tube into the other (defuelling) machine.

There are 18 CANDU reactors in operation in Canada, 16 of them in Ontario at the Pickering, Bruce and Darlington sites. Two additional units at Bruce that have been in long-term layup for technical reasons may now be restarted, although large-scale maintenance to repair the plant's steam generators will be required before the restart can go ahead. The remaining Canadian CANDUs are at Gentilly in Quebec and Point Lepreau in New Brunswick.

Elsewhere, there are four CANDU reactors in operation at Wolsung in South Korea and two at Qinshan in China. Under Nikolai Ceaucescu, Romania had a large nuclear programme and had started civil works on five CANDU reactors at a site called Cernavoda. Work restarted on the first reactor at that site in the mid-1990s and the first reactor on site went into operation in 1996. Work has now restarted on the site's second reactor.

2.4.2 Pressurised heavy-water reactor

In the late 1960s and early 1970s Canada supplied CANDU reactors to both Pakistan and India. Those two countries declined to sign the nuclear non-proliferation treaty in 1968 and although Pakistan's reactor at Karachi was under international safeguards, the rest of Pakistan's nuclear programme was not and as a result Canada broke off its nuclear relationship with Pakistan in 1974. Canada's nuclear cooperation with India also ended in 1976, after India's detonation of a nuclear test in 1974 and after discussions over improved safeguards failed. In the 1990s a tentative relationship was re-established for safety reasons via the CANDU Owners Group, an international information exchange for plant operators.

Karachi's nuclear power station remained Pakistan's only reactor for many years until, in the 1990s, it revived its nuclear power programme, this time in the form of a PWR supplied by China. India, however, continued its own development of power reactors and eventually built a series of pressurised heavy-water reactors (PHWRs) based on the CANDU design. It now has 13 PHWRs in operation and five further units under construction.

2.4.3 Steam-generating heavy-water reactor

A prototype reactor designed by the UK AEA and built (but now shut down) at its Winfrith site in Dorset was intended to combine the features of the CANDU reactor and PWR in a steam-generating heavy-water reactor (SGHWR).

In the SGHWR a calandria filled with heavy-water moderator surrounds vertical pressure tubes; in each pressure tube is a single 4 m fuel array, which is a cluster of zircaloy-clad enriched uranium dioxide fuel pins through which ordinary light-water coolant is circulated. The light water is allowed to boil as it passes through the core, as happens in the BWR. The steam is circulated directly through turbine generators, before being condensed and pumped again from below into the pressure tubes.

As in the CANDU reactor, the SGHWR power level is varied by varying the level of the moderator and the SGHWR can be scrammed by pouring boron solution into interstitial tubes in the calandria. As fission products begin to build up when the fuel ages the amount of boron in the moderator is reduced to compensate.

Like other light-water designs the SGHWR has a comparatively high power density. The SGHWR design is unusually flexible in size: the combination of highly efficient heavy-water moderator and pressure tubes design makes it feasible to build the SGHWR in a smaller size than other designs. However, the design was not advanced past the demonstration reactor at the UKAEA's site at Winfrith in Dorset, which in fact produced more power than the early Magnox designs, being rated at 100 MWe. The Winfrith reactor began operating in 1967 and was eventually shut down in 1990.

2.5 Fast breeder reactors

All the reactors described above rely primarily on the energy released by fission of uranium-235 by slow neutrons. These are known as 'thermal' neutrons, and the reactors are often referred to as thermal reactors. In a thermal reactor only a small proportion of the neutrons are absorbed by uranium-238, to make fissile plutonium-239. Materials like uranium-238, which can be converted, by absorption of neutrons, into a fissile material like plutonium-239, are called 'fertile' (thorium-232 is also fertile; it can be converted by neutron bombardment into fissile uranium-233).

The comparison between fissile nuclei consumed and fertile nuclei converted to fissile is called the 'conversion ratio'. In a burner reactor the conversion ratio is always less than 1, but it is also possible to design a reactor that produces more fissile nuclei than it consumes, and this is known as a 'breeder' reactor.

In some designs, the fuel assemblies in the fissile core are surrounded by a 'blanket' layer of assemblies containing the fertile materials. Depleted uranium can be used as a fertile material so the fast reactor offers the ability to greatly increase the amount of energy produced from each tonne of mined uranium – it is estimated that the amount of energy produced from each tonne can be increased sixty-fold.

The core of a fast reactor can be extremely compact, as no moderator is required, so an extremely efficient coolant is required. As a result fast reactors may be cooled by liquid metal (sodium, in most cases so far) and are therefore sometimes called liquid metal-cooled fast reactors (LMFRs).

2.5.1 Early designs

As it happens, the Experimental Breeder Reactor EBR-l, the first reactor ever to power electric generating equipment, was a fast breeder reactor. However, the first real power-generating breeder reactors, on a small scale, came into operation in the 1960s. The first were Britain's demonstration fast reactors, a 15 MWe reactor that started up in 1962 at Dounreay in Scotland, and the US's EBR-2, a 20 MWe reactor that went into operation in 1964 and was not shut down until 30 years later. In the UK, DFR was followed by a much larger prototype fast reactor (PFR), which was rated at 270 MWe and started up in 1974. That too operated until 1994.

The PFR marked the end of the UK's large-scale independent fast reactor programme; thereafter work was focused on thermal reactors, although the UK remains involved in international research and development of fast reactors.

Russian fast reactors came slightly later. The first, a 12 MWe unit known as BOR-60 in Dimitrovgrad, was started up in 1969–70, and it was followed ten years later by a full-scale reactor in Beloyarsk, Ekaterinburg rated at 600 MWe, and known as BN-600. Both Russian units are still in operation. Between these projects a reactor was built at Aktau (then in the USSR, now in Kazakhstan) around 1973. This former weapons-grade plutonium-producing reactor is now being decommissioned with assistance from the UK and US governments as part of a non-proliferation programme of assistance to Kazakhstan.

Japan's immense investment in its nuclear industry also included a fast breeder programme. Japan's prototype fast reactor, Monju, took ten years to construct and went into operation in 1994. It is rated at 280 MWe and was due to be followed by a larger demonstration fast reactor, which would be rated at 660 MWe. However, a general reining in of Japan's nuclear programme and the long process involved in bringing Monju up to power have put its successor on hold.

France had taken the lead in fast reactor development by the 1970s and 1980s, reflecting its commitment to a large nuclear industry. It had started up a prototype fast reactor known as Phénix in 1973 that operated successfully and was still in operation

Figure 2.5 Monju fast breeder reactor, Japan [Nuclear Fuel and Power Reactor Development Corporation (PNC)]

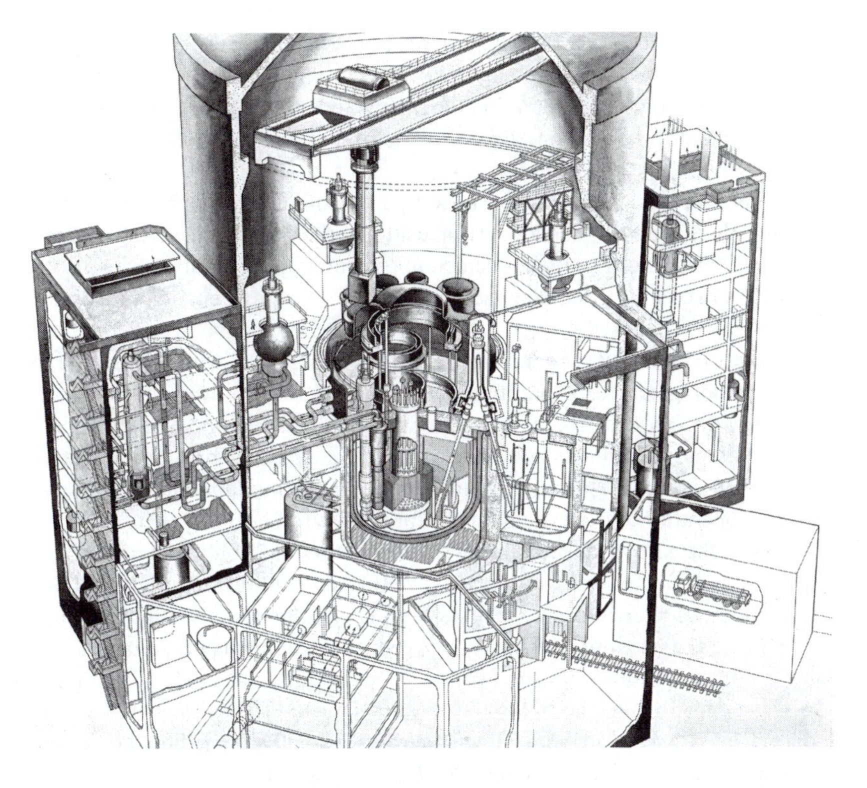

Figure 2.6 Cutaway of the Superphénix fast breeder reactor at Creys Malville, France [EDF/NERSA]

in 2005. The success of Phénix led France to scale up its design to produce the first commercial-scale fast reactor, a 1,200 MWe plant known as Superphénix, whose construction began in 1977 and which was started up in 1985.

2.5.2 Superphénix

Superphénix was regarded as a commercial industrial prototype, whose initial objectives were the production of electricity and plutonium. However, from its entry into operation up until 1994, the station suffered a variety of technical problems and it seldom operated. A major issue was the refuelling machinery; a secondary vessel also filled with liquid sodium that was used to transfer fuel in and out of the reactor core. Leaks in the refuelling machine eventually meant it had to be abandoned and an alternative refuelling strategy developed.

On 3 July 1990, the reactor was shut down because of impurities in the sodium in the core. This shutdown lasted so long that the plant's owner was forced to relaunch the administrative procedure for an authorisation to operate – a procedure that became necessary after two years of not operating. The new authorisation decree, dated 11 July 1994, authorised Superphénix to operate as a research and demonstration reactor. Instead of operating as a commercial prototype power reactor, the revised programme, which was to cover the 1995–2000 period, had two new objectives. They were evaluation of the operation of a fast breeder reactor as a 'net consumer of plutonium' and study of the possibility of destroying long-lived radioactive wastes in fast breeder reactors.

However, in February 1997 the Conseil d'Etat annulled the 1994 decree. Subsequently the plant operator filed a new request for authorisation to operate the reactor. In February 1998 the government decided not to grant this request and the plant was shut down permanently. The plant had operated for the equivalent of 174 days at full power.

Work on FBRs lost its urgency in the 1970s and 1980s, when it became clear that far from being scarce, as had first been thought, uranium was a fairly common resource, and in any case the high usage rates that had been predicted in the first expansion of the nuclear industry had been dramatically scaled back as the industry's growth slowed. It became clear that there was sufficient uranium fuel for the reactors currently in operation and for the industry for many years.

However, the further development of fast reactors has continued as an international effort, initially focusing on revised safety and economic requirements for the next generation of nuclear power plants. Work is also continuing on improving fuel burnup and the fuel recycling technology to reduce the amounts of radioactive waste produced at plants, and wider use of passive systems for further safety. Designs currently on the table include the BN-800M in Russia, the DFBR in Japan and the PFBR in India. There is also an advanced LMFR design in small and medium power range, developed by General Electric in the USA (see Chapter 8).

In recent years there has been some revival of interest in the long-term future of the fast breeder and a number of different types of LMFRs are being studied as so-called Generation IV reactors, likely to be required in the second part of the century (see Chapter 9).

2.6 New generations

The early reactors developed in 1950–60s are generally referred to as 'first generation' reactors and only a few are still running today. The reactors described above are referred to as the second generation, and they make up the bulk of the reactors currently in operation.

Some operating reactors represent an advance on Generation II designs: the advanced BWR (ABWR) already in operation in Japan is referred to as Generation III, and this generation also encompasses the so-called advanced reactors such as the European Pressurised Water Reactor (EPR). These focus on improving safety, economics and severe accident management scenarios. More than a dozen (Generation III) advanced reactor designs are at various stages of development. Some have evolved from the PWR, BWR and CANDU designs, while others are more radical and may be designated Generation IIIa. The best-known radical new design is the Pebble Bed Modular Reactor or HTR, which uses helium as coolant, at very

Figure 2.7 Superphénix nuclear power plant at Creys-Malville, France [IAEA Image Bank]

high temperatures, to drive a turbine directly. Generation III designs are currently on top of the list for new potential reactor projects and are described in more detail in Chapter 8.

Beyond this, there are a number of more advanced designs on the drawing board that incorporate more complex fuel and operating cycles. This group of reactors is being designed for a future scenario when uranium is less readily available and the need to develop or 'breed' new fuels is more urgent. This group of reactors is under design and development for potential deployment around the middle of the century and are collectively referred to as Generation IV designs. The group is described in Chapter 9.

Chapter 3
Nuclear safety and three major accidents

Nuclear reactor safety requires that three functions should be fulfilled at all times:

- The chain reaction must be controlled and so must the power generated.
- The fuel must be cooled during and after operation, including after the chain reaction has stopped, as residual heat remains in the reactor core caused by continued radioactive disintegrations and fissions.
- Radioactive products must be retained and controlled.

The safety philosophy relies upon two main principles: three protective barriers and so-called defence in depth.

The three protective barriers are intended to contain radioactivity in all circumstances. The first barrier is the fuel, inside which most of the radioactive products are already trapped. The fuel is contained within a metal cladding (magnox, steel or zircaloy) which presents a barrier to stop the products escaping.

The second barrier is the reactor coolant system, housed within a containment enclosure which includes the reactor vessel containing the core constituted by the fuel within its cladding. For the third barrier the reactor coolant system is also enclosed in a containment including a biological shield made of very thick concrete.

For the European PWR now under construction in Finland, for example, this construction is a double shell resting upon a thick basemat, whose inner wall is covered with a leak-tight metal liner.

3.1 Defence in depth

The concept of 'defence in depth' is defined by the International Atomic Energy Agency as 'a hierarchical deployment of different levels of equipment and procedures in order to maintain the effectiveness of physical barriers placed between a radiation source or radioactive materials and workers, members of the public or the environment, in operational states and, for some barriers, in accident conditions'.

The objectives of this approach are:

- To compensate for potential human and component failures;
- To maintain the effectiveness of the barriers by averting damage to the facility and to the barriers themselves and
- To protect the public and the environment from harm in the event that these barriers are not fully effective.

Defence in depth can also refer to the application of more than one protective measure for a given safety objective, such that the objective is achieved even if one of the protective measures fails.

The IAEA's International Nuclear Safety Advisory Group (INSAG – see below) defines five levels of defence in depth:

- Level 1: prevention of abnormal operation and failures;
- Level 2: control of abnormal operation and detection of failures;
- Level 3: control of accidents within the design basis;
- Level 4: control of severe plant conditions, including prevention of accident progression and mitigation of the consequences of severe accidents and
- Level 5: mitigation of radiological consequences of significant releases of radioactive materials.

The levels of defence are sometimes grouped into three safety layers: hardware, software and management control.

This concept translates into recent reactor design, as with the EPR. According to its designers, the reactor provides defence in depth via

- safe design, quality workmanship and diligent operation, with incorporation of the lessons of experience feedback in order to prevent occurrence of failures,
- providing a means of surveillance for detecting any anomaly leading to departure from normal service conditions in order to anticipate failures or to detect them as soon as they occur,
- providing the means of action for mitigating the consequences of failures and preventing core meltdown; this level includes use of redundant systems to automatically bring the reactor to safe shutdown; the most important of these systems is the automatic shutdown by the insertion of control rods into the core, which stops the nuclear reaction in a few seconds; in addition, a set of safeguard systems, also redundant, are implemented to ensure the containment of the radioactive products.

In the reactor's 'design basis', potential failures are anticipated and methods and systems included to deal with or mitigate the consequences of the fault. For example, in a PWR a steam generator tube break is an accident which, if it occurs, leads to a transfer of water and pressure from the primary system to the secondary system. The primary side pressure drop automatically induces a reactor shutdown. If the pressure continues to fall, it triggers an automatic injection of water into the reactor vessel.

In the EPR, a safety injection system lower than the set pressure of the secondary system safety valves prevents the steam generators from filling up with water in such a case. This has a dual advantage: it avoids the production of liquid releases and considerably reduces the risk of a secondary safety valve locking in an open position.

The defence in depth approach goes further than such 'design basis accidents': it postulates the failure of all these three levels, resulting in a 'severe accident' situation, and aims to provide the means of minimising the consequences of such an accident, both by mitigation systems and by training operators to act in a way that will minimise the consequences of accidents.

This safety philosophy has been both developed in response to, and tested by, several accidents in nuclear reactors. The three most important, and whose effects were felt most widely, were at Windscale in the UK, Three Mile Island in the US and Chernobyl in Ukraine (but at that time part of the USSR). The following description of these three events is largely taken from 'Reactor Accidents', by David Mosey, published by *Nuclear Engineering International* magazine.

3.2 Windscale, 7 October 1957

The Windscale Piles 1 and 2, completed in late 1950, were plutonium production reactors constructed as part of Britain's nuclear weapons programme (see Chapter 2).

Graphite moderated and forced-air cooled, the reactors were originally designed to use natural uranium metal fuel. However, after a short period of operation very slightly enriched uranium (0.73 per cent ^{235}U) was introduced in 1953. The level of enrichment was chosen to give the minimum production cost for plutonium. A later report noted that this change meant that the piles' 'heat rating went well above the design figures'.

At the time the Windscale piles were designed, knowledge of the effects of radiation on graphite was limited. It was understood that graphite expanded under irradiation but in 1949, data on the amount of growth were conflicting. When the Windscale Piles were started up, weekly measurements of graphite growth were taken, but no operational plans had been made to deal with the resulting stored energy. This stored energy, inherent in the changing structure of the graphite, is known as Wigner energy.

In September 1952 a spontaneous release of Wigner energy occurred while the Pile was shut down. As a result of this experience, a method was instituted for the controlled release of Wigner energy at regular intervals and thermocouples were installed to monitor the graphite temperature.

The method adopted to release the energy consisted of shutting off the cooling air flow, bringing the reactor to criticality and allowing the graphite temperature to rise until it reached the point at which the Wigner release started. The reactor was then shut down again while the energy was released.

Wigner release intervals were originally set at 20,000 MWd (megawatt days, a measure of the amount of energy that the reactor had produced), subsequently extended to 30,000 MWd. In view of what seemed to be the increasing difficulty of obtaining successful releases, in 1957 consideration was given to a further significant increase to 50,000 MWd intervals. However, until more experience had been gained a 40,000 MWd interval was selected, implying a Wigner release operation for October of that year.

3.2.1 Accident sequence

On 7 October 1957 Pile No. 1 was shut down at just after 01.00 in preparation for the planned Wigner release operation. Preparatory work included checking (and where necessary replacing) the uranium and graphite thermocouples used for following the Wigner release, switching off the shutdown cooling fans and opening roof inspection-hole covers and a door in the base of the stack to minimise cooling airflow through the pile.

About 18 hours after shutdown (19.25) the reactor was made critical and the power level was gradually raised to an indicated 1.8 MW by 01.00 on 8 October. This would be a low reading since with the upper horizontal control rods fully inserted (as was the case for this operation) the ion chamber measuring reactor power would be masked. However, since for Wigner release operations power was controlled on the basis of temperature readings, the only effect of this masking would be to alter the relation between power changes and movement of the lower horizontal control rods.

At about this time a fuel cartridge temperature of 250 °C was indicated in two channels. Since this was laid down as an upper limit for the initial stages of a Wigner release, the control rods were inserted and the reactor was shut down by 04.00. Initially, the graphite temperatures rose as expected, in the manner characteristic of a Wigner release, but by about 09.00 (8 October) the physicist in charge felt that the graphite temperatures seemed to be falling rather than rising and further heating was needed if the release was to be completed.

Second nuclear heatings had been required during three previous Wigner energy releases in 1954, 1955 and 1956 but had not been applied until 24 hours after the last temperature rise and when all graphite temperatures were observed to be falling. In this case, however, the second heating was initiated at 11.05. Uranium temperatures rose sharply, the highest thermocouple indication being 380 °C about 15 min after restart. Power was reduced and the thermocouple reading fell to 334 °C within 10 min. The pile was maintained at this lower power level until 17.00 at which time it was shut down.

In the course of the following day (9 October) graphite temperatures showed considerable variation but with a general tendency to increase. For example, at one location (channel group 20/53) the temperature rose steadily from an indicated 255 °C at the time of the start of the second heating phase to an indicated 405 °C at 22.00, some 36 hours later.

Because of this high temperature trend the chimney base door and inspection holes were closed at 21.00. This would allow the natural chimney draught to induce a cooling flow of air through the pile. However, the cooling effect was not considered adequate and at 22.15 the fan dampers were opened for 15 min to provide a positive flow of air. This operation was repeated at 00.01 on 10 October (for 10 min), at 02.15 (for 13 min) and at 05.10 (for 30 min) and resulted in a fall in all graphite temperatures save the highest (the 20/53 channel group), where the temperature rise was merely halted.

At the end of the fourth damper opening (at 05.40) the stack activity monitor showed a sharp increase, which was attributed to the first movement of air through the pile. Over the next two and a half hours this reading fell steadily, as would be expected, but then it began to rise again. The graphite temperature in channel group 20/53 continued to rise, and at 12.00 a temperature of 428 °C was recorded, prompting further damper openings at 12.10 (for 15 min) and 13.40 (for 5 min). During these openings a second, and very much larger, increase in stack activity was recorded and at about the same time high activity readings on a nearby roof (the Meteorological Station) were reported. From this it was inferred that one or more fuel failures had occurred and at 13.15 the shutdown cooling fans were switched on to drive air through the pile so that the channel scanning equipment could be used to identify the location of the failed fuel. The scanner, however, proved to be jammed – a situation experienced during previous Wigner energy releases, and probably attributable to overheating.

Sampling of the exhaust air from the pile revealed high levels of particulate activity and suggested that serious fuel failure must have occurred, necessitating rapid identification of the affected channel and discharge of the fuel. Channel group 21/53 had shown very rapid temperature increases and, by 16.30, had an indicated temperature of about 450 °C.

Since the scanning gear was not working this channel was inspected visually and the fuel inside was seen to be glowing. Initial attempts to discharge the fuel were fruitless and consideration was given to blanking off the channel group using graphite plugs. However, shortly afterwards (about 17.00) it was found that around the affected group of channels a total of about 150 channels (40 groups) were at red heat. The graphite was effectively on fire.

Two expedients were then adopted: discharge of fuel from hot channels and discharge of fuel from a ring of channels surrounding the hot channels in order to create a firebreak. This latter was successfully accomplished and two more rows of channels above the hot region were subsequently discharged as the fire threatened to spread upwards. Attempts were made to extinguish the fire using carbon dioxide, but these were unsuccessful, and successive observations through an inspection hole revealed a steady growth in the fire. A glow at the rear face of the pile was evident at 18.45. At 20.00, yellow flames were seen and half an hour later the flames were blue.

At this point the use of water was first considered, even though it presented two potential hazards. First there was the danger of a hydrogen–oxygen explosion which could blow out the stack filters and lead to a large and uncontrolled release of radioactivity and second, a possible criticality hazard through the introduction of water into an air-cooled pile. These dangers, however, had to be balanced against the distinct possibility of the whole pile catching fire should the graphite temperature rise much higher than 1,200 °C.

By midnight it was decided that if all other measures failed to achieve a temperature reduction, then water should be used. The fire brigade was ordered to stand by and arrangements were made to enable water to be injected into the discharged channels.

At 01.38 the graphite in channel group 20/53 near the top of the high-temperature area had an indicated temperature of 1,000 °C and optical pyrometer readings in this area indicated a fuel temperature of about 1,300 °C. Strenuous efforts over the next two hours were successful in dislodging burning fuel elements from the top row of channels in the 'hot' area, but by that time the fire was too widespread for such action to have very much effect. At 04.00 blue flames were still in evidence and the graphite appeared to be burning. Efforts to discharge the burning fuel continued, but by 07.00 it was clear that the fire was not being controlled and it was decided to flood the pile. After all the site workers had taken cover, water injection started at 08.55.

For the first hour there was little observable effect and flames were still visible, but after the shutdown cooling fans were turned off at 10.10 the fire immediately began to subside. Water injection continued until 15.10 on 12 October by which time the pile was cold.

3.2.2 Immediate causes of the accident

In their report to the chairman of the United Kingdom Atomic Energy Authority, a Committee of Inquiry chaired by Sir William Penny concluded that the fire began during the second nuclear heating which was applied too soon and too rapidly. Most probably the rapid rise of temperature during the second nuclear heating caused the failure of one or more of the aluminium fuel element cans. The exposed uranium oxidised, providing an additional heat source which, in combination with the heating effect of the later Wigner releases, started the fire.

The report drew attention to the fact that contrary to the pile physicist's impression that the initial nuclear heating had not triggered a Wigner release, examination of the thermocouple traces showed that a substantial number of graphite thermocouple readings showed steadily increasing temperatures. In previous Wigner release operations, where application of a second nuclear heating had been made, this had not begun until virtually all graphite temperatures were observed to be falling. The second heating was also applied at an unusually rapid rate – the recorded rate of uranium temperature rise was 10 °C/min, in contrast with the 2 °C/min rate adhered to under normal operational practice.

In addition, the positioning of thermocouples in the pile was such that misleading uranium temperature indications would be given. Since in Wigner release operations the control rods were adjusted to give a flux peak 0.9 m closer to the front of the pile than under normal operation and airflow through the pile was minimised, maximum uranium temperatures would occur about 2.1 m closer to the front of the pile than the thermocouples. This would result in maximum uranium temperatures being about 40 per cent higher than the thermocouple readings.

Calculations suggested that, at the position of peak neutron flux, uranium temperatures would have risen rapidly from an initial value of 340 °C to as high as 450 °C in the course of the second nuclear heating and would have remained at this level for some minutes.

Based on the known behaviour of the Windscale fuel elements and their irradiation history, the report concluded that this treatment would have brought about immediate clad failure for fuel in this region.

That the fuel failures were not diagnosed earlier is attributed to the nature of the air flow through the pile. The initial attempt to induce cooling air flow through the pile at 21.00 on 9 October involved closing the chimney base door and the inspection holes and opening the inlet air dampers. At this point the air in the pile would be stagnant and that in the chimney would be cool, so there would be little differential pressure available to drive air through the core and carry fission products up the stack to the monitors. The subsequent second and third damper openings were too short in duration to establish air flow. But at 05.10 on 10 October the fourth opening of the dampers for a 50 min period was sufficient to drive gaseous fission products and particulates up the chimney, giving rise to the sharp increase in the stack monitor reading at 05.40. However at the time this was attributed to the first arrival at the monitors of air that had been resident in the core for some time and hence would carry a larger concentration of radioactive material. The operations to establish air flow through the core naturally caused an accelerated rate of oxidation of the uranium fuel, while switching on the shutdown cooling fans at 13.45 (following the very large stack monitor readings noted at 05.40) caused a further rapid increase so that by 15.00 an intense fire was burning in the region of the 20/53 channel group.

The initiating events of the Windscale fire could be described as operator error on the part of the pile physicist exacerbated by inadequate instrumentation of the pile. The thermocouples were not correctly positioned for monitoring maximum uranium temperatures during a Wigner energy release and the conditions in the core (i.e., stagnant air) precluded early detection of the damaged fuel elements. However, the official report was at pains to point out that there appeared to exist no form of operating manual for the pile physicist's guidance. The only piece of documentation available appeared to be a memorandum dated 14 November 1955 suggesting actions to follow at three different temperatures during a Wigner release.

All other details of pile operation either existed as committee minutes or seemed to be traditions (habitual practice) without any form of written authority at all. The lack of formal documentation was to some extent a consequence of the Windscale operations being part of a 'pioneering and urgent' programme. However, there were also deficiencies and inadequacies in the organisation. It was unclear how responsibilities were divided between the various branches of the organisation, and communication between various groups was poor. In some cases this meant that groups were unaware of technical changes made by another or unaware of the significance of those changes for their own work.

There was 'undue reliance on technical direction from the committee'. This correlated with responses to other incidents elsewhere and the conclusion that a single body must have responsibility for safety and a 'line organisation' should replace the committee.

The operating staff were also not well supported by technical advice: the Windscale Works Technical Committee, for example, had never examined the records from recent Wigner releases.

3.3 Three Mile Island, 28 March 1978

The Unit 1 reactor at Three Mile Island was a Babcock and Wilcox designed PWR operated by Metropolitan Edison. The unit achieved first criticality on 28 March 1978. One year later on 28 March 1979, the reactor core was destroyed when, following a total loss of feedwater, core cooling was seriously impaired. The accident is also described in detail in a fact sheet produced by the US Nuclear Regulatory Commission that is available in its 'electronic reading room' (http://www.nrc.gov/reading-rm/doc-collections/fact-sheets/3mile-isle.html).

3.3.1 Plant description

The reactor core comprised a 3.27 m diameter by 3.65 m high cylinder made up of 177 fuel assemblies, contained in a 4.35 m diameter by 12 m high carbon steel pressure vessel. Each fuel assembly was a 15 × 15 array containing 208 fuel rods plus spaces for control and instrumentation elements. The fuel was zircaloy-4 clad uranium dioxide enriched to 2.57 per cent ^{235}U.

The reactor's primary coolant circuit comprised two loops, each with two main circulating pumps and a single vertical once-through steam generator. Primary coolant operating pressure was 2,150 psi with an outlet temperature of 319.4 °C. A pressuriser connected to one of the loops controlled the primary system pressure and provided surge volume to accommodate expansion and contraction of primary coolant water. The pressuriser was connected to a reactor coolant drain tank via a power-operated relief valve (PORV). A block valve was provided upstream of the relief valve and could be remotely operated in the event the relief valve stuck open or leaked.

Under normal operating conditions a primary coolant let-down and make-up system removed 45–70 USgpm (US gallons per minute) from the primary system, purified and cooled it and returned it to the primary system through one of the three high-pressure make-up pumps. Make-up supply was automatically controlled by pressuriser level signals. Two of the make-up pumps formed part of the emergency coolant injection system. When actuated each pump injected borated water into all four reactor cooling inlet pipes. Under injection conditions the let-down flow was automatically stopped.

Feedwater flow was from the condensate pumps through eight parallel resin beds (condensate polishers) to condensate booster pumps, through low-pressure feedwater heaters to the main feedwater pumps (one per steam generator) and through the high-pressure feedwater heaters to the steam generators. The main feedwater pumps were backed up by two electric (dc) pumps (one per steam generator) and a single, shared, steam turbine driven pump. On loss of the main feedwater pumps these three auxiliary pumps would start automatically and provide a direct feed to the steam generators.

The valves linking these pumps to the steam generators were arranged to open when the steam generator level fell to 76.2 cm or less.

3.3.2 Accident sequence

At 04.00 on 28 March the reactor was working at 98 per cent of full power. At this moment all the in-service condensate polishers isolated automatically as a result of water entering an instrument air line through a check valve which had stuck open. This interruption in condensate flow immediately caused all the condensate booster pumps and main feedwater pumps to trip. Alarms sounded within the control room.

Because heat was no longer being transferred to the secondary loop, water pressure and water temperature in the primary loop rose. This was normal and caused the operators no concern.

Within three seconds the primary system pressure had risen to the pressuriser relief valve setpoint. The PORV (pressure relief valve) opened automatically, releasing steam into a holding tank. Backup pumps within the secondary loop automatically turned on but they were disconnected from the system by cutoff valves, unknown to the operators. The reactor tripped automatically but heat was still generated.

At 12 s, primary system pressure had fallen to 2,205 psig, at which the pressuriser relief valve should have closed. It did not do so, although the PORV indicator light in the control room went out, apparently indicating that the valve was now closed. The valve was actually still open and continued to release steam and water, creating a loss of coolant accident (LOCA).

The emergency injection water (EIW) was activated, and water flowed into the primary loop. This device was designed to keep the water at a safe level in the event of a LOCA. Operators were not too concerned by this, as the EIW had turned itself on many times in the past when there had been no leak.

As coolant flowed from the core through the pressuriser, the instruments available to reactor operators provided confusing information. There was no instrument that showed the level of coolant in the core. Instead, the operators judged the level of water in the core by the level in the pressuriser, and since it was high, they assumed that the core was properly covered with coolant. In addition, there was no clear signal that the PORV was open. As a result, as alarms rang and warning lights flashed, the operators did not realise that the plant was experiencing a loss of coolant accident. They took a series of actions that made conditions worse by simply reducing the flow of coolant through the core.

At 4 min operators observed that the water level in the primary system was rising and that the pressure was decreasing and turned off the EIW. The water level still appeared to be rising but was actually dropping. The water, along with the steam, was now being released through the PORV.

At 8 min an operator noticed that the valves for the backup pumps in the secondary loop were off and opened them. The secondary side was now operating normally.

By 15 min after the start of the event, approximately 3,000 gallons of water had escaped from the primary loop. The instrument that checks radioactivity levels did not trigger an alarm, so operators still had no reason to suspect a LOCA.

At 45 min gauges in the control room wrongly indicated that the water level was up, but in fact the water level in the primary loop continued to drop.

At 1 hour and 20 min water was boiling in the primary circuit and as a result steam, not water, was passing through the pumps that drive water through the primary loop. This caused them to begin to shake violently. Two of the four pumps were turned off. Twenty minutes later the other two pumps went off. This meant that steam within the primary loop, now no longer circulating with the water, rose. Because water was not circulating, the core heated up even more, converting more of the water into steam.

Two hours and 15 min after the first valve trip, the top of the core was no longer covered by water, as it had been converted into steam. The control rods, in turn, reacted to the superheated steam and began to release hydrogen and radioactive gases. These were also released through the PORV.

An operator from the next shift, coming on duty, noticed that the PORV discharge temperature was abnormally high and stopped the leak by shutting the PORV's backup valve. More than a quarter of a million gallons of radioactive cooling water had been discharged since the PORV first opened.

Operators still did not realise that the water level in the primary loop was low. The water within the loop continued to boil away, which caused more damage to the core, more heat and more radioactivity.

The operators received the first indication that radiation levels were going up at 2 hours and 30 min, and 15 min later radiation alarms sounded. The core was now half-uncovered and the radioactivity of the water in the primary loop was 350 times its normal level.

A general emergency was declared at 3 hours. Some operators by now believed the core was uncovered, but others thought the temperature readings were wrong. Operators pumped water into the primary loop but pressure remained high and a backup valve was opened to lower the pressure.

One hour 30 min later the hydrogen released by the control rods exploded within the containment structure, causing a pressure spike on the control room gauges and an audible thud.

After 15 hours the primary loop's pumps were turned on, which circulate water around the core. The core's temperature was finally brought under control, although half of it was melted and part of it disintegrated.

Because adequate cooling was not available, the nuclear fuel overheated to the point at which the zirconium cladding (the long metal tubes which hold the nuclear fuel pellets) ruptured and the fuel pellets began to melt. It was later found that about one-half of the core had melted during the early stages of the accident.

By the evening of 28 March, the core appeared to be adequately cooled and the reactor appeared to be stable. But new concerns arose by the morning of Friday 30 March. A significant release of radiation from the plant's auxiliary building, performed to relieve pressure on the primary system and avoid curtailing the flow of coolant to the core, caused a great deal of confusion and consternation. In an atmosphere of growing uncertainty about the condition of the plant, the governor of Pennsylvania, Richard L Thornburgh, consulted with the NRC about evacuating the population near the plant. Eventually, he and the NRC Chairman, Joseph Hendrie, agreed that it would be prudent for those members of society most vulnerable to radiation to evacuate the area. Thornburgh announced that he was advising pregnant women and pre-school age children within a 5-mile radius of the plant to leave the area.

The presence of a large hydrogen bubble in the dome of the pressure vessel stirred new worries. The concern was that the hydrogen bubble might burn or even explode and rupture the pressure vessel. In that event, the core would fall into the containment building and perhaps cause a breach of containment. The hydrogen bubble was a source of intense scrutiny and great anxiety, both among government authorities and the population, throughout the day on 31 March. The crisis ended when experts determined on Sunday 1 April that the bubble could not burn or explode because of the absence of oxygen in the pressure vessel. Further, by that time, the utility had succeeded in greatly reducing the size of the bubble.

3.3.3 Health effects

Detailed studies of the radiological consequences of the accident have been conducted by the NRC, the Environmental Protection Agency, the Department of Health, Education and Welfare (now Health and Human Services), the Department of Energy and the State of Pennsylvania. Several independent studies have also been conducted. Estimates are that the average dose to about 2 million people in the area was about 1 millirem (similar to a chest x-ray), in comparison to the natural radioactive background dose of about 100–125 millirem per year for the area. The maximum dose to a person at the site boundary would have been less than 100 millirem.

In the months following the accident, although questions were raised about possible adverse effects from radiation on human, animal and plant life in the TMI area, none could be directly correlated to the accident. Thousands of environmental samples of air, water, milk, vegetation, soil and foodstuffs were collected by various groups monitoring the area. Very low levels of radionuclides could be attributed to releases from the accident. However, most of the radiation was contained.

The accident was caused by a combination of personnel error, design deficiencies and component failure. There is no doubt that the accident at Three Mile Island permanently changed both the nuclear industry and the NRC. Public fear and distrust increased, NRC's regulations and oversight became broader and more robust and management of the plants was scrutinised more carefully.

Among the major changes which have occurred since the accident:

- Upgrading and strengthening of plant design and equipment requirements. This includes fire protection, piping systems, auxiliary feedwater systems, containment building isolation, reliability of individual components (pressure relief valves and electrical circuit breakers) and the ability of plants to shut down automatically.
- Identifying human performance as a critical part of plant safety, revamping operator training and staffing requirements, followed by improved instrumentation and controls for operating the plant and establishment of fitness-for-duty programmes for plant workers to guard against alcohol or drug abuse.
- Improved instruction to avoid the confusing signals that plagued operations during the accident.

- Enhancement of emergency preparedness to include immediate NRC notification requirements for plant events and an NRC operations centre which is now staffed 24 h a day. Drills and response plans are now tested by licensees several times a year, and state and local agencies participate in drills with the Federal Emergency Management Agency and NRC.
- Establishment of a programme to integrate NRC observations, findings and conclusions about licensee performance and management effectiveness into a periodic, public report.
- Regular analysis of plant performance by senior NRC managers who identify those plants needing additional regulatory attention.
- Expansion of NRC's resident inspector programme whereby at least two inspectors live nearby and work exclusively at each plant in the US to provide daily surveillance of licensee adherence to NRC regulations.
- Expansion of performance-oriented as well as safety-oriented inspections and the use of risk assessment to identify vulnerabilities of any plant to severe accidents.
- Strengthening and reorganisation of enforcement as a separate office within the NRC.
- The establishment of the Institute of Nuclear Power Operations (INPO) and formation of what is now the Nuclear Energy Institute to provide a unified industry approach to generic nuclear regulatory issues and interaction with NRC and other government agencies.
- Additional equipment by licensees to mitigate accident conditions and monitor radiation levels and plant status.
- Expansion of NRC's international activities to share enhanced knowledge of nuclear safety with other countries in a number of important technical areas.

3.3.4 Current status

Today, the TMI-2 reactor is permanently shut down and defuelled, with the reactor coolant system drained, the radioactive water decontaminated and evaporated, radioactive waste shipped off-site to an appropriate disposal site, reactor fuel and core debris shipped off-site to a Department of Energy facility and the remainder of the site being monitored. The owner says it will keep the facility in long-term, monitored storage until the operating licence for the TMI-1 plant expires at which time both plants will be decommissioned.

3.4 Chernobyl, 26 April 1986

The RBMK reactor is a vertically orientated graphite-moderated direct cycle (boiling-water) pressure tube reactor (see Chapter 2). The core is a 12 m diameter by 7 m high cylinder built of graphite blocks, threaded by almost 1,700 vertical zirconium tubes which contain the reactor fuel and the various control and shut-off rods. The reactor fuel is zirconium-clad uranium oxide enriched to 2 per cent ^{235}U.

The heat transport system comprises two cooling loops, each with four main circulating pumps and two steam drum/separators. The discharge from the pumps

Figure 3.1 The Chernobyl reactor immediately after the accident

goes to a large 900 mm diameter header, which feeds a series of 300 mm diameter distribution headers which serve groups of feeder pipes. As the coolant flows up the fuel channels it boils and the steam/water mixture is fed to the steam drum, and thence to two 500 MWe turbines. Coolant flow through the core is matched to power levels by large throttling valves on the main coolant pump discharges. In addition, at the point of each feeder connection to a distribution header there is an individual control valve which is used to regulate the flow through the individual channels to maximise the margin to fuel dryout under a variety of operating conditions, while keeping the channel power as high as possible.

The reactor core is contained in a steel vessel which is filled with a mixture of helium and nitrogen to maintain an inert atmosphere for the hot graphite (which is at about 600 °C) and to promote heat transfer. An arrangement of graphite rings on the pressure tubes ensures good thermal contact between the tubes and the moderator to provide a heat transfer path from the graphite via the pressure tubes to the coolant.

Reactor control is via 211 boron carbide absorber rods of which 24 are 'safety' or 'scram' rods, which are usually fully outside the core to provide reactivity depth on shutdown. There are also 12 'local automatic regulation' rods, which maintain power shape using signals from four lateral ionisation chambers, and 12 'automatic power regulation' rods which maintain total reactor power and are arranged in three sets of four ganged rods. The 139 'manual control' rods are manipulated by the operators in response to changing reactor conditions to keep the automatically controlled rods within their range of travel. Finally there are 24 'bottom entry' short rods used for manual axial power shaping. Auxiliary absorbers are also installed temporarily in a new core to hold down reactivity. There are 2,409 of these boron steel absorbers initially, which are progressively replaced by fuel as burnup increases.

The reactor and most of the coolant circuit are contained within a number of thick-walled, leak-tight enclosures that are linked to two pressure suppression pools ('bubbler ponds'). However, the upper sections of the fuel channels and the steam drums are not enclosed. The top of the reactor is covered with a 3 m thick steel and concrete shielding slab through which the fuel channels pass. A single fuelling machine above the reactor is used for refuelling.

A significant feature of the RBMK reactor design is the influence of reactor coolant conditions on the power coefficient at various power levels. At higher power levels the positive reactivity effect of an increase in coolant void is more than offset by the negative reactivity effects of an increase in fuel temperature, so power rises are self-limiting. This is not the case at power levels below about 20 per cent of full power and the coolant void is the dominant influence on the power coefficient, so at low power levels the reactor is unstable and difficult to control. For this reason a fundamental operating procedure forbids sustained operation at power levels below 20 per cent.

The maximum rate at which the control rods could be inserted into the reactor was 400 mm per second, implying an insertion time of about 15 s for a rod moving from the fully withdrawn position. As a result operators were required to maintain a 'reactivity margin' of at least 30 rods inserted at least 1.2 m into the core at all times.

3.4.1 The test

The Chernobyl reactivity excursion was triggered by a test aimed at improving capabilities for coping with loss-of-power emergencies.

The idea was to show that a new voltage regulation arrangement would allow that turbine generator inertia was enough to power the emergency core cooling system for enough time (35 s) to start the emergency diesels. The test should have been done during initial startup, but it had been postponed because the startup was on New Year's Eve. The plant had therefore entered service in breach of safety requirements.

In the test planned for 25–26 April 1986, rather than using the emergency core cooling system (ECCS) pump itself, the load of the ECCS pump was to be simulated by running four main coolant pumps from the coasting down turbine generator. The test was to be performed at a level of 2–32 per cent of power, as the unit was reducing power for a routine maintenance shutdown.

At 01.00 on 25 April power reduction began. It reached 50 per cent power at 13.05 and one of the two turbines was disconnected. The load required in-house was transferred to the second turbine. At this time the grid controller in Kiev requested that generation continue, to compensate for the loss of a thermal power station elsewhere on the network. The reactor was held at 50 per cent power for 9 hours, when the reduction was resumed.

The ECCS had been disconnected at 14.00 and remained so.

At 00.28 on 26 April, with reduction continuing, a mistake in setting the control rods caused power to fall to less than 30 MWt – 1 per cent of full power – well below the 700 MWt minimum specified in operating procedures. The power fell rapidly and uncontrollably because boiling in the core reduced.

Raising power from this point was difficult because the core still contained lots of xenon gas (^{139}Xe) – a radioactive product that decays over a few hours but while it exists is a strong 'poison' for the fission reaction. The fission process was effectively 'smothered' by the xenon and it took until 01.00 on 26 April to raise the power to 200 MWt (7 per cent), which was achieved despite the xenon poisoning by withdrawing a large number of control rods from the core – against operating procedures, which specified a minimum of 30 rods remaining in the core. Eventually just 6–8 rods were still inserted. The reactor then operated well below the 20 per cent minimum specified. This made it difficult to keep tabs on parameters such as the water temperature, flow, levels and voiding across the core.

When there were just a few control rods in the core, it made it more difficult to respond to an emergency situation and shut the reactor down quickly. Also, with fewer absorbers in the core, the positive void coefficient in the core was still larger. The effect was a feedback loop: the increase in power caused the reactor power to increase faster.

The power profile across the core was not consistent, because there were areas of xenon poisoning. The power was greatest at the top and bottom of the reactor. At the bottom of the reactor this would magnify the effects of voiding (boiling) in the channel inlets – an area where the scram rods slowly inserted from the top of the core would be least effective.

Despite the fact that the reactor was operating far outside acceptable parameters the test was continued. As a result, at 01.03 and 01.07 on 26 April additional coolant pumps were started up. But the low power level meant the pumps drove the coolant through the core much faster than expected. There was a sharp reduction in steam production, so pressure fell and the water level in the steam drum/separators also fell. Water in the coolant circuits was very close to boiling but did not boil, a situation which left the reactor very sensitive to boiling if it were to occur. The operators tried unsuccessfully to throttle the coolant flow.

The low water levels in the steam drum/separator would be expected to trip the reactor. To avoid this, the operators disconnected the protection system, along with a second trip system that would have shut down the reactor as system pressure passed its setpoints.

The operators then increased the feedwater flow to try to correct the low pressure in the steam drum/separator. Instead, it reduced steam production still further, along with reactivity. The operator removed more control rods to compensate.

The operator decided at 01.22 that the steam drum/separator level was high enough and reduced feedwater flow. This meant that warmer water began to arrive at the channel inlets at the bottom of the core and began to boil. At 01.23 the operators decided that the reactor was returning to stability and decided to complete the test.

The turbine stop valve was closed so it would run down. This would normally have tripped the reactor, since the other turbine was shut down, but the operators had already blocked this emergency trip function so that the test could be re-run if necessary.

Thirty seconds after 01.23, reactor power began to increase as steam began to be produced. Because of the reactor's positive void coefficient, the fatal power excursion began.

The shift manager ordered a full emergency shutdown just ten seconds later, apparently in response to the power increase. But the reactor became prompt critical four seconds later at 01.23.44, and power surged to over 100 times the normal full power. The cause of the power surge was the subject of some dispute.

Reactivity in the bottom of the core, where reactivity was already very high, would actually have been increased by the insertion of the control rods. In the RBMK design water acts as a neutron absorber rather than as a moderator (which increases reactivity). But the boron control rods had a graphite 'follower' attached at the bottom, and graphite is a moderator. When the control rods were introduced the graphite (moderator) displaced water (absorber) and sped up the reaction at the bottom of the core.

It is also possible that cavitation (sudden boiling of the water) in the pumps caused them to fail and xenon began to burn out so its poison effect ended.

Shocks were felt and the operator noticed the control rods were not inserting fully. Power was cut so they would fall under gravity. But fuel in the core fragmented, leading to massive boiling of coolant water and a second power surge to around 440 times the full power. At 01.23.48 the steam that had built up beneath the top shield blew it off, rupturing all the pressure tubes and destroying the reactor. Seconds later more explosions, thought to be due to hydrogen and carbon monoxide, were seen, along with fires on the top of the reactor and reactor building.

3.4.2 Examining the accident

In 'Reactor Accidents', David Mosey points out that there was no equipment failure that led to the accident. In addition, only one operator action could have been described as human error as all the others, although they were violations of safety rules, were taken deliberately. He says the accident stemmed principally from a series of actions taken by those manning the plant. It attained catastrophic proportions because of design features of the RBMK reactor type. However, underlying both the operational and design aspects of the accident were deep-seated organisational and institutional problems in the Soviet nuclear industry.

The RBMK reactor was particularly sensitive to operator malpractice and human failings in general. Examples were the positive void coefficient, the containment philosophy, the slow rate of emergency control rod insertion, the absence of a fast shutdown system and the ease with which protective systems could be disabled. Too much reliance was placed on operators to keep their reactors out of trouble.

Meanwhile, the accident revealed weakness and malpractice throughout the industry. The defects in the reactor protection system were known but it had not been redesigned. In construction, production considerations overrode quality concerns. The attitude to nuclear safety of the operators was poor, and in addition their operating instruction and documentation made it difficult to complete procedures properly. It had become standard practice to break operating rules.

The principal institutional failures were that:

- The organisational structure was such that safety responsibility was collective rather than individual.
- Existing hazards had been identified but not acted on.
- Concerns about production and output overrode safety and quality considerations. The test programme was drawn up and implemented without any concern for review or approvals.
- The safety envelope of the equipment was not well understood by those managing the operation of the plant. The plant was overseen by an electrical engineer who was not a reactor specialist.

3.4.3 After-effects of the accident

The apparent effects of the Chernobyl accident were summarised in a digest report, 'Chernobyl's Legacy: Health, Environmental and Socio-Economic Impacts', released by the Chernobyl Forum in September 2005. The Forum is made up of eight UN specialist agencies, including the International Atomic Energy Agency (IAEA), World Health Organization (WHO), United Nations Development Programme (UNDP), Food and Agriculture Organization (FAO), United Nations Environment Programme (UNEP), United Nations Office for the Coordination of Humanitarian Affairs (UNOCHA), United Nations Scientific Committee on the Effects of Atomic Radiation (UNSCEAR) and the World Bank, as well as the governments of Belarus, Russia and Ukraine. The digest assesses the 20-year impact of the accident.

The report said that the accident had major health consequences, especially for thousands of workers exposed in the early days, who received very high radiation doses, and for the thousands more stricken with thyroid cancer. However, it did not find 'profound negative health impacts' on the rest of the population in surrounding areas nor widespread contamination outside the existing restricted areas that would continue to pose a substantial threat to human health.

Among the findings:

- Approximately 1,000 on-site reactor staff and emergency workers were heavily exposed to high-level radiation on the first day of the accident. Among the more than 200,000 emergency and recovery operation workers exposed during the period from 1986 to 1987, an estimated 2,200 radiation-caused deaths can be expected. This includes some 50 emergency workers who died of acute radiation syndrome. The estimated 4,000 casualties may occur during the lifetime of about 600,000 people under consideration, including those from within restricted zones. As about a quarter of people die from spontaneous cancer not caused by Chernobyl radiation, the radiation-induced increase of only about 3 per cent will be difficult to observe. However, in the most exposed cohorts of emergency and recovery operation workers some increase in particular cancer forms (e.g. leukaemia) in particular time periods has already been observed.

- An estimated five million people currently live in areas of Belarus, Russia and Ukraine that are contaminated with radionuclides due to the accident. For the majority, exposures are within the recommended dose limit for the general public but about 100,000 of them live in areas classified in the past by government authorities as areas of 'strict control' and they receive a higher dose. Remediation of those areas and application of some agricultural countermeasures continue. Further reduction of exposure levels will be slow, but most exposure from the accident has already occurred. The existing 'zoning' definitions need to be revisited and relaxed in the light of the new findings.
- About 4,000 cases of thyroid cancer, mainly in children and adolescents at the time of the accident, have resulted from the accident's contamination. There have been nine deaths but all others have recovered. Otherwise, the team of international experts found no evidence for any increases in the incidence of leukaemia and cancer among affected residents.
- Most emergency workers and people living in contaminated areas received relatively low whole-body radiation doses, comparable to natural background levels. Because of the relatively low doses to residents of contaminated territories, no evidence or likelihood of decreased fertility has been seen among males or females. Also, because the doses were so low, there was no evidence of any effect on the number of stillbirths, adverse pregnancy outcomes, delivery complications or overall health of children. A modest but steady increase in reported congenital malformations in both contaminated and uncontaminated areas of Belarus appears related to better reporting, not radiation.
- Relocation proved a 'deeply traumatic experience' for some 350,000 people moved out of the affected areas. Although 116,000 were moved from the most heavily impacted area immediately after the accident, later relocations did little to reduce radiation exposure.

Alongside radiation-induced deaths and diseases, the report labels the mental health impact of Chernobyl as 'the largest public health problem created by the accident' and partially attributes this damaging psychological impact to a lack of accurate information. These problems manifest as negative self-assessments of health, belief in a shortened life expectancy, lack of initiative and dependence on assistance from the state. Persistent myths and misperceptions about the threat of radiation have resulted in 'paralyzing fatalism' among residents of affected areas.

Poverty, 'lifestyle' diseases now rampant in the former Soviet Union, and mental health problems pose a far greater threat to local communities than radiation exposure.

3.4.4 Radiological contamination

Most of the strontium and plutonium isotopes were deposited within 100 km of the damaged reactor. Radioactive iodine, of great concern after the accident, has a

Figure 3.2 Chernobyl nuclear power plant showing the sarcophagus in 2001, Ukraine [Vadim Mouchkin/IAEA Image Bank]

short half-life and has now decayed away. Strontium and caesium, with a longer half-life of 30 years, persist and will remain a concern for decades to come. Although plutonium isotopes and americium-241 will persist, perhaps for thousands of years, their contribution to human exposure is low.

3.4.5 Remediation on the reactor site

The reactor is protected by a hastily built 'shelter' sometimes known as the 'sarcophagus'. It was erected quickly, using existing structural elements such as the remaining reactor building, which made it difficult to fully assess the stability of the damaged unit.

Some structural parts of the shelter have corroded in the past two decades. The main potential hazard posed by the shelter is the possible collapse of its top structures and the release of radioactive dust. The unstable structures have been strengthened in recent years, and construction of a New Safe Confinement covering the existing shelter, which should serve for more than 100 years, starts in the near future. The new cover will allow dismantlement of the current shelter, removal of the radioactive fuel mass from the damaged unit and an eventual decommissioning of the damaged reactor.

A comprehensive strategy still has to be developed for dealing with the high-level and long-lived radioactive waste from past remediation activities. Much of this waste was placed in temporary storage in trenches and landfills that do not meet current waste safety requirements.

Panel 3.1 International Nuclear Event Scale

The International Atomic Energy Agency has introduced a simple scale that is designed for communicating to the public in consistent terms the safety significance of events at nuclear facilities. Informally, on this scale 'incident' is used to denote an event that has been halted by the 'defence in depth' systems in the plant. 'Accident' refers to an event where all those systems have been degraded.

Level 0 (deviation): an event with no safety significance.
Level 1 (anomaly): an event beyond the authorised operating regime, but not involving significant failures in safety provisions, significant spread of contamination or overexposure of workers.
Level 2 (incident): an event involving significant failure in safety provisions, but with sufficient defence in depth remaining to cope with additional failures, and/or resulting in a dose to a worker exceeding a statutory dose limit and/or leading to the presence of activity in on-site areas not expected by design and which require corrective action.
Level 3 (serious incident): a near accident, where only the last layer of defence in depth remained operational, and/or involving severe spread of contamination on-site or deterministic effects to a worker and/or a very small release of radioactive material off-site (i.e. critical group dose of the order of tenths of a millisievert).
Level 4 (accident without significant off-site risk): an accident involving significant damage to the installation (e.g. partial core melt) and/or overexposure of one or more workers resulting in a high probability of death and/or an off-site release such that the critical group dose is of the order of a few millisieverts.
Level 5 (accident with off-site risk): an accident resulting in severe damage to the installation and/or an off-site release of activity radiologically equivalent to hundreds or thousands of terabecquerels of 131I, likely to result in partial implementation of countermeasures covered by emergency plans, for example, the 1979 accident at Three Mile Island, USA (severe damage to the installation), or the 1957 accident at Windscale, UK (severe damage to the installation and significant off-site release).
Level 6 (serious accident): an accident involving a significant release of radioactive material and likely to require full implementation of planned countermeasures but less severe than a major accident, for example, the 1957 accident at Kyshtym, USSR (now in Russian Federation).
Level 7 (major accident): an accident involving a major release of radioactive material with widespread health and environmental effects, for example, the 1986 accident at Chernobyl, USSR (now in Ukraine).

Chapter 4

Operating experience

At the time of the Three Mile Island accident in 1979 the nuclear industry had a lot to learn about the safe operation of its plants, as the accident revealed.

It was equally important to improve the plants' economic performance. One measure of performance is the unit capability factor, which is the percentage of its maximum energy generation that a plant is capable of supplying to the electrical grid, and it is limited by factors within the plant management control, including unplanned shutdowns.

In 1980, the year following the Three Mile accident, the US industry's median capability factor was just 62.7 per cent – slightly higher for the US's PWRs than for BWRs – indicating that the plants were unavailable to generate power for nearly 40 per cent of the time. The situation was similar in other countries. Industry magazine *Nuclear Engineering International*, which has tracked load factors for many years, placed the UK at the bottom of its 'load factor league table' in 1981, with an average of 43 per cent, while the US did slightly better at 54 per cent (NEI's load figures are slightly lower than other measures as it measures against the design rating, whereas some plants were started up with a slightly lower stated rating). Germany averaged 57 per cent, Japan 60 per cent, France and Switzerland 65 per cent, Canada (expected to be the best performer on this measure, as its plants refuel online so no refuelling outages are required) at 69 per cent and Sweden at 71 per cent.

The US response to the accident at Three Mile Island was the establishment of the Institute of Nuclear Power Operations (INPO) by the US nuclear industry in late 1979. INPO's mission is 'to promote the highest levels of safety and reliability in the operation of nuclear electric generating plants'. The nuclear utility industry leaders established INPO as an independent organisation – independent from governmental agencies and independent from any individual member.

Each of the 34 utilities in the US with operational nuclear plants is a member of the Institute, while non-US nuclear utility organisations participate in the Institute's International Programme. Ten nuclear steam system suppliers and architect-engineering and construction firms worldwide involved in nuclear work also participate in INPO through a supplier programme.

At the same time nuclear operators have exchanged information and experience on an international basis as part of 'Owners Groups' – the BWR Owners Group, Westinghouse Owners Group, CANDU Owners Group and so on. This enables operators to develop solutions for technical issues and discuss new developments and ageing issues directly with the plant designers and vendors.

The INPO model was extended worldwide in response to the Chernobyl accident, when the World Association of Nuclear Operators was set up, headquartered in London and with regional centres in Europe, Russia, Asia and the US (effectively the INPO organisation).

The key technical activities of these groups can be divided into four cornerstone programmes. They are as follows:

- Evaluations – Periodic evaluations are conducted for each operating nuclear electric plant in this country.
- Training and Accreditation – Training programmes for key personnel at each plant are accredited by the independent National Nuclear Accrediting Board.
- Events Analysis and Information Exchange – INPO analyses operating experience and feeds back lessons learned to the industry.
- Assistance – This includes plant visits, courses, seminars and workshops.

INPO's evaluation programme cornerstone is a direct response to a post-TMI recommendation that '... the industry must ... set and police its own standards of excellence to ensure the effective management and safe operation of nuclear electric generating plants'.

Teams of qualified and experienced personnel conduct these evaluations, focusing on plant safety and reliability. The evaluation teams are augmented by senior reactor operators, other peer evaluators from operating units similar to those at the station being evaluated and host utility peer evaluators. The scope of the evaluation includes traditional functional categories such as operations, maintenance and engineering that generally correspond to the nuclear station organisation. The areas evaluated include organisational effectiveness, operations, maintenance, engineering, radiological protection, chemistry and training.

In addition, the teams evaluate cross-functional performance areas – processes and behaviours that cross organisational boundaries and that address organisational integration and interfaces. The cross-functional evaluation includes areas such as safety culture, self-assessment and corrective action (learning organisation), operating experience, human performance and training.

The performance of operations and training personnel during simulator exercises is included as part of each evaluation. Also included, where practicable, are observations of plant startups, shutdowns and major planned evaluations. Evaluations of each operating nuclear station are conducted at an average interval of 21 months. Results from more than 875 plant evaluations INPO has conducted to date show substantial improvements in the conduct of plant operations, enhanced maintenance practices and improvements in equipment and human performance.

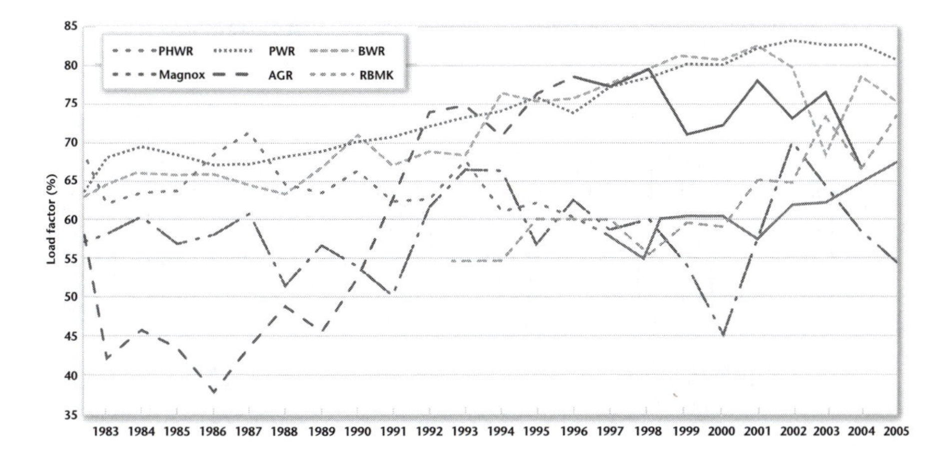

Figure 4.1 *Variation of load factor for various reactor types over time [published with permission of* Nuclear Engineering International*]*

During the 1980s electric utilities knew that for nuclear power to be commercially viable, operating and maintenance costs had to be reduced. One way to do this was to improve plant utilisation, measured by comparing the number of hours each reactor spends on-line to its theoretical maximum. This measure can vary slightly depending on the way in which the maximum number of hours is defined, for example whether it includes programmed shutdowns or periods when the reactor is 'available to operate' but not called on. Hence the utilisation may be variously referred to as plant capability, availability or load factor; the trend of these figures for an individual plant should be similar although the numbers may differ.

4.1 Improving plant performance

INPO's annual 'performance indicators' show how safety and performance are necessarily closely linked and have been improved in step. The ten performance indicators are:

- unplanned capability loss factor
- fuel reliability
- thermal performance
- chemistry performance
- collective radiation exposure (by reactor type, PWR or BWR)
- volume of solid radioactive waste produced (by reactor type, PWR or BWR)
- unit capability factor
- unplanned automatic scrams
- industrial safety accident rate
- safety system performance

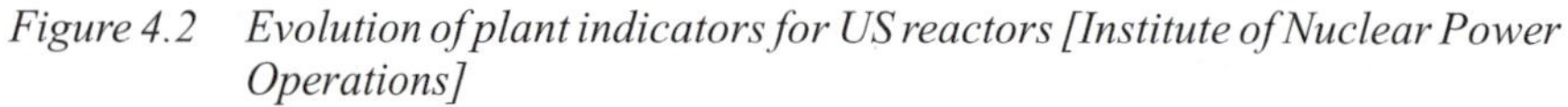

Figure 4.2 Evolution of plant indicators for US reactors [Institute of Nuclear Power Operations]

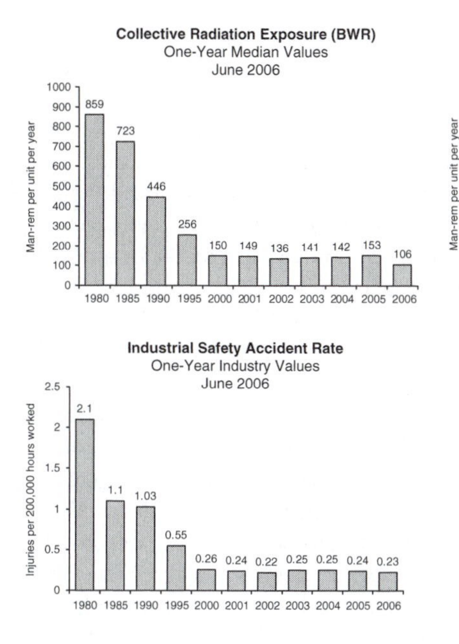

Figure 4.2 Continued

The inclusion of indicators on industrial accidents and waste generated illustrates how the concept of safety has grown since INPO was started in 1980. The idea was that technological 'defence in depth' should be underpinned by a 'safety culture' in the operating organisation that extends from the reactor and its operators, through the 'balance of plant' or conventional parts of the plant and operators of those plant sectors, to the maintenance and management teams and extending to the entire company. Industrial accident rates, radiation exposure and radioactive waste produced are good indicators of whether all staffing levels are operating to the right levels of safety and performance.

It should be obvious: after all the effects of accidents or problems in the non-nuclear part for the plant can be dramatic. For example, as with all industrial plants, fire is a major hazard, especially in the turbine hall where hydrogen is used as a lubricant and coolant. Fire safety has moved on from March 1975, when one unit at the Browns Ferry plant in Alabama, USA, was shut down after a fire. Two electricians had been working with an open candle flame and a highly flammable caulking material, polyurethane. The caulking material ignited and the resulting fire, while it did not directly affect the nuclear part of the plant, did enough damage to ensure it was uneconomic to restore it and bring it back to power for 25 years – a process that is now finally under way.

Fire hazards in the turbine hall have proved terminal for two other reactors: Vandellos in Spain and the second unit at Chernobyl, which had remained in action after the accident at the fourth site. In both cases the so-called 'reactor island' was untouched by the fire and although it was a significant degradation of the 'defence in depth' strategy, core cooling was not impaired. But in both cases the cost of rebuilding the turbine hall was insurmountable. In the mid-1990s when India suffered more than one fire in the turbine halls of its PHWRs there was again no impact on the nuclear island, but a programme to change its turbine design and carry out upgrades at all its nuclear stations meant all were shut down for long periods and reduced the fleet's annual load factors to 20 per cent in the worst-affected year.

A series of institutional changes have helped improve capacity factors since then. New fuel designs permitted higher burnups. Such improvements permitted the expansion of periods between refuelling outages to be increased from 12 to 18 months and sometimes to two years. Refuelling outages have also been cut from as much as three months in the past to about a month today. Methods of undertaking other maintenance and capital replacement during these outages or even during operations have also been improved. Time requirements for planned and unplanned maintenance have been shortened. Some changes have simply allowed best-practices to be captured and replicated. For example, Palo Verde is the largest nuclear station in the US. But maintenance and refuelling outages at its three reactors could be over 100 days long. One answer was to change the maintenance teams. In the past, the three reactors had been operated as entirely separate entities, each of which had its own maintenance, operations and management team, and there was no coordination between the units. Instead, the maintenance engineers were assigned to teams with associated support, and each team took responsibility for a single plant system or group of systems. The teams could then quickly build expertise on their own systems as they rotated around the plant. Each team was also incentivised to suggest technical or organisational improvements. Palo Verde's typical outage dropped dramatically, so that even the longest annual outage (a once per decade inspection) should not take more than 50 days. In Europe, outages were reduced still more effectively: some Scandinavian units expect now to complete a refuelling outage in 12–14 days and new reactor designs aim at reducing that still further.

More recently techniques such as risk-informed maintenance have also been used. Using this maintenance strategy reliability does not depend on regular shutdown and visual inspection of components and equipment that may include dismantling and reassembling the equipment. Instead, the equipment is fitted with sensors and monitors that allow its condition to be assessed on a day-to-day basis. The benefit is twofold: if the wear pattern is within limits the equipment does not need to be shut down for maintenance for a longer period; but if an abnormal or dangerous situation is diagnosed an early maintenance shutdown or replacement can be scheduled, before the equipment reaches the point of breakdown. Monitoring equipment can include temperature and vibration measurements, acoustic monitoring etc.

It would be difficult to decide whether longer fuel cycles, shorter outages, better maintenance or the holistic approach to managing and maintaining the plants have been more important than the other factors in improving performance, though the reduced outage time is a major component. Because refuelling and maintenance outages must still continue at reactors, they are clearly approaching a technical limit for average plant capacity factors. Improved nuclear power performance in the future must come from other sources.

This holistic approach to safety and performance has had a dramatic effect, which has been more marked in some countries than others. The UK's lifetime figures have always lagged behind, because its AGR design presented technical problems in loading fuel that had to be solved for each reactor, and the problem was not really tackled until the industry went through privatisation in the 1990s (see Chapter 6). Despite some good years at the end of the century when figures were above 80 per cent, its lifetime figure is 53.9 per cent. But according to NEI's Load Factor League Table, average load factors in 2005 were 93 per cent in Finland, 70.4 per cent in France, 86.3 per cent in Germany, 65 per cent in Japan, 83.4 per cent in Switzerland and 84.9 per cent in the USA.

By INPO's measure (capability factor) the US has seen reactors increase their effectiveness in operation to around 90 per cent. That means that although nominal nuclear generating capacity has remained roughly constant from 1990, the amount of electricity produced has increased by around a third because capacity utilisation has increased that much. The increase in nuclear power generation due to capacity factor increases is roughly equivalent to building a number of new power plants operating at former capacity levels.

As performance has improved, power production costs have been reduced accordingly. Production costs for US nuclear power plant operation and maintenance, plus fuel costs, averaged 1.8 cents per kilowatt-hour by the turn of the century, according to INPO.

4.2 Ageing and maintenance

As plant performance has improved, the plants themselves have been ageing, and as a result a series of new issues have arisen that have to be dealt with.

For nuclear plant owners this means implementing 'life cycle management', a long-term plan for maintaining the plant's systems, structures and components in good working order.

Within the life cycle management plan preventive maintenance consists of routine, scheduled activities to keep a plant's safety as well as non-safety equipment running or capable of functioning if needed. With more than 35 years of experience, plant operators have learned how systems wear and can refurbish or replace the vast majority before they fail.

Corrective maintenance is performed on equipment that fails routine testing, breaks down during operation or does not perform adequately. When the operation of an important component degrades or fails, plant operators conduct detailed, root-cause analyses, take corrective action and share the lessons learned with all other plant operators throughout the industry and with regulators.

All nuclear plant owners participate in a comprehensive, industry-wide materials management programme that identifies when components of nuclear plants need to be inspected, repaired and replaced to ensure that the materials from which they are made perform according to standards. This integrated, coordinated approach is intended to

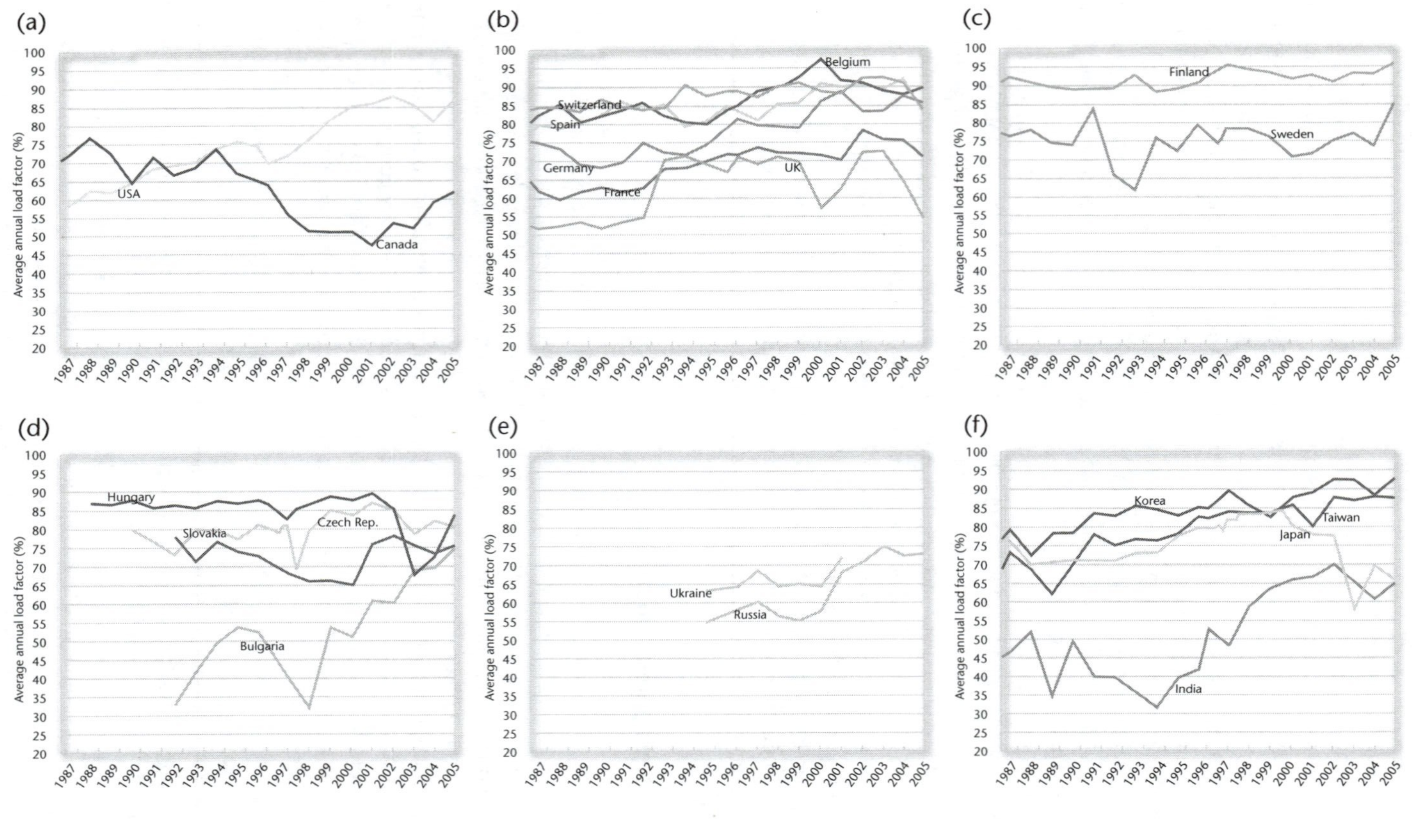

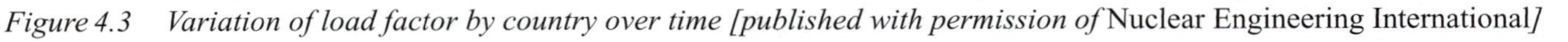

Figure 4.3 Variation of load factor by country over time [published with permission of Nuclear Engineering International*]*

prevent component malfunction due to materials degradation occurring during plant operation. The programme involves an assessment of materials degradation, impacts on key systems and components and a process for promptly communicating results throughout the industry.

Physical ageing can affect structures, systems and components due to physical, chemical and/or biological processes. Examples of physical ageing include wear, heat or radiation damage and corrosion. The term material ageing is also used.

Non-physical ageing processes include the process of becoming out-of-date (i.e. obsolete) owing to the evolution of knowledge and technology and the associated changes in codes and standards. Examples of non-physical ageing include unavailability of qualified spare parts for old equipment, incompatibility between old and new equipment and outdated procedures or documentation (e.g. which do not comply with current regulations). Strictly, this is not always ageing as defined above, because it is sometimes not due to changes in the structure, system or component itself. Nevertheless, the effects on protection and safety, and the solutions that need to be adopted, are often very similar to those for physical ageing; the management of non-physical ageing is therefore often addressed within the same programme as that for the management of physical ageing. The term technological obsolescence is also used.

The oldest nuclear reactors in operation are now more than 40 years old and many are expected to carry on operating for a good many years beyond that. The 40-year lifetime, which had been nominally attached to most reactors since the 1960s, has very little technical basis but is in effect an accounting lifetime.

Up until the end of the 1980s, when most reactors now operating were being constructed, it was assumed that this time limit was real and it was also assumed that most of the major components – the reactor pressure vessel, steam generators, turbine generator, major pipework and so on – would remain in operation for the lifetime of the plant.

In some reactor designs there are technical life-limiting factors. The graphite cores and concrete pressure vessels used at the UK's Magnox and advanced gas reactors, for example, are subject to ageing stresses as the reactors operate and cannot be replaced. For some other types of reactor it has been relatively straightforward to replace major components, even if it was originally assumed they would last the life of the plant. PWRs are a good example, perhaps because they are based on a compact and modular design used in submarines.

Components can be removed and factory-fabricated replacements installed, while leaving the concrete containment and other civil structures largely untouched.

In some cases this has happened because ageing effects in the reactor components have made major repairs or replacements necessary (such as for PWR steam generators) and in others (such as instrumentation and control systems) because it has become economically efficient to replace the original equipment.

4.2.1 Steam generator replacement

The fate of the PWR steam generators illustrates how much has been learnt about the long-term management of the components of nuclear power plants.

PWRs have two, three or four coolant loops through the core, each with a steam generator that transfers the heat from the superheated high-pressure water that has travelled through the reactor core, to a secondary circuit where water is turned into the high-pressure steam that will turn the turbine. They are the largest components in the primary circuit: the steam generators of the EPR being built at Olkiluoto are 24 m high (twice as large as the reactor pressure vessel, which is 12 m high) and weigh 500 tonnes. To transfer heat as efficiently as possible inside the steam generators the surface area separating the water in the primary and secondary circuits must be maximised. To do this, water piped from the core is split into thousands of 'u-tubes' a couple of centimetres in diameter. The u-tubes (5,980 in each steam generator at Olkiluoto) are so-called because they reach from the primary water entry at the base up the full 24 m height of the steam generator and back down to the primary water exit point, also at the bottom. The secondary side water surrounds the u-tubes.

The steam generator design includes a series of spacers and restraints that stop the u-tubes from vibrating as the water passes through them.

A few years after the first PWRs went into operation it became clear that the steam generators represented a major maintenance challenge. Primary circuit water at high temperatures contained impurities that could precipitate out as the water temperature dropped and leave deposits inside the u-tubes (especially in the 'u' itself, the least accessible part of the 24 m tubes). The restraints did not always completely eliminate vibration problems. And as time went by the steel of the u-tubes could undergo corrosion, or its granular structure could gradually change to one that was more brittle and thus more subject to cracking.

Cracking in the u-tubes would allow primary circuit water to leak into the secondary circuit, potentially carrying fission products that would introduce activity into the turbine circuit and increase the exposure of workers in that 'non-nuclear' part of the plant. What was more, such leaks were clearly a degradation of the 'defence in depth' philosophy of the plant. The US Nuclear Regulatory Commission says it places a high priority on ensuring that steam generator tube degradation is carefully monitored. To obtain an operating licence, applicants must show that the consequences of a steam generator tube rupture would not exceed the NRC's conservative limits for radiation doses offsite or outside the plant (described in the agency's regulations in Title 10 of the Code of Federal Regulations, Part 100). Plant operators also are required to have emergency procedures for mitigating steam generator tube ruptures and leaks.

Once a plant is in service, its operator is required to inspect and repair or remove from use all tubes found to contain flaws exceeding certain limits. There are also operational leakage limits to ensure that if any of the tubes leak, the plant will be shut down quickly.

Plant operators carried out regular and intensive steam generator inspections. To do this, new types of equipment had to be developed. These included tiny cameras that could be inserted into the tubes to give visual evidence of cracking or deposits, vibration monitors and so-called Hall-effect probes that were inserted into the u-tubes and used small magnetic fields to identify areas where 'discontinuities' in the metal properties might indicate the beginnings of cracking.

At the same time, techniques for flushing the u-tubes to remove buildup of corrosion products (known as crud) were developed. Some relied on high-pressure water jets to dislodge the buildup. This 'water lancing' could be employed at every outage. Other techniques include chemical solutions to try to dissolve crud buildup. This type of chemical cleaning had to be carefully planned to ensure that all the reagents and dislodged crud could be entirely removed from the steam generator before it was refilled ready for the unit to restart.

In addition to these inspections and remedial measures, careful control of the chemistry in the primary circuit was employed to keep crud deposition and the steel ageing process to a minimum. This usually entailed dosing the primary circuit water with reagents to ensure the acid–alkali balance was more alkali (elevating the pH). This was not a simple equation: the chemical balance also had to be suitable for the core and other parts of the primary circuit, and there were other chemical additions required, for example, varying amounts of boron according to conditions in the core.

The NRC describes the progress of steam generator tube management in the US as follows.

During the early-to-mid-1970s, when all plants, except one, had mill-annealed Alloy 600 steam generator tubes, thinning of the mill-annealed Alloy 600 steam generator tube walls due to the chemistry of the water flowing around them was the dominant cause of tube degradation. However, all plants have changed their water chemistry control programmes since then, virtually eliminating the problem with tube thinning.

After tube thinning, tube denting became a primary concern in the mid-to-late-1970s. Denting results from the corrosion of the carbon steel support plates and the buildup of corrosion product in the crevices between tubes and the tube support plates. Measures have been taken to control denting, including changes in the chemistry of the secondary (i.e. non-radioactive) side of the plant. But other phenomena continue to cause tube cracking in plants with mill-annealed Alloy 600 tubes.

The extensive tube degradation at pressurised-water reactors (PWRs) with mill-annealed Alloy 600 steam generator tubes resulted in tube leaks, tube ruptures and midcycle steam generator tube inspections. This degradation also contributed to the permanent shutdown of other plants.

Inspection, cleaning and careful chemistry control could minimise the problem, but cracking in the u-tubes continued to be a problem. It could be solved by lining tubes that had indications of cracking with a metal liner that was fused to the inside of the tubes concerned. Ultimately, however, badly affected tubes had to be 'plugged' so that primary water did not enter the tube and the primary/secondary circuit barrier was maintained.

Dealing with steam generator cracking had prompted development of devices that could detect cracking indications just a few millimetres long embedded up to 20 m inside a curved narrow tube. However, at some plants slow but steady plugging rates meant that the performance of the steam generators – and hence the plant efficiency and economics – was becoming permanently degraded. By the start of the 1990s some plants had steam generator plugging rates of up to 20 per cent, with concomitant effects on the plant output. Now the minutiae of inspection and control had to be supplemented by the most dramatic of large-scale maintenance procedures: replacing the entire steam generator.

It was not clear at first whether this was a feasible procedure. The huge steam generators were the largest components within the reactor containment and had been expected to last the lifetime of the plant. To replace them, the thick pipework of the primary and secondary circuits would have to be cut. The steam generator would have to be removed from the containment and replacements brought in and erected. In some cases the containment would have to be cut away to provide an exit point for the steam generator, although at some reactors there was an 'equipment hatch'.

The first steam generator replacement was performed in the USA, at the Surrey plant. Three steam generators were replaced at unit 1 in 1979 and three at unit 2 the following year. In order to complete the replacement, unit 1 was in outage for 303 days; work at unit 2 took nearly 100 days fewer, clocking in at 209 days.

Over the next ten years, replacements like this became almost routine. The steam generator connections were cut and the steam generator winched down to lie horizontal on a specially built vehicle. The steam generator was wheeled through an existing equipment hatch – generally with only inches to spare – or specially cut access way. The process was repeated with the remaining steam generators. Replacement steam generators entered by the reverse route and were winched into place and the pipework welded together. The time required for the operation dropped dramatically over the next few years. By 1996, France's Electricité de France and the reactor vendor Framatome were able to replace the three steam generators at Gravelines 2 in 33 days. The radioactive exposure incurred during the procedure was cut just as dramatically: from 7.14 manSv and 5.86 manSv per steam generator at Surrey 1 and Surrey 2, to 0.46 manSv at Gravelines.

As the mill-annealed Alloy 600 steam generator tubes began exhibiting degradation in the early 1970s, the industry pursued improvements in the design of future steam generators to reduce the likelihood of corrosion. In the late 1970s, Alloy 600 tubes were subjected to a high-temperature thermal treatment to improve the tubes' resistance to corrosion. This thermal treatment process was first used on tubes installed in replacement steam generators put into service in the early 1980s and no significant degradation problems have been observed in plants. Plants which have replaced their steam generators since 1989 have primarily used tubes fabricated from thermally treated Alloy 690, which is believed to be even more corrosion-resistant than thermally treated Alloy 600. Thermally treated Alloy 690 is presently used in the steam generators at 27 plants.

Most of the newer steam generators, including all of the replacement steam generators, have features that make the tubes less susceptible to corrosion-related damage. These include using stainless steel tube support plates to minimise the likelihood of denting and new fabrication techniques to minimise mechanical stress on tubes.

4.3 Vessel closures

Apart from the steam generators, a series of other ageing behaviours have had to be remedied on each nuclear reactor design.

A weakness in the PWR is the 'penetrations' in the head of the reactor pressure vessel. These tubes, lined with stainless steel alloys, allow connections for core internals such as the control rod drives or instrumentation channels to pass through the vessel head. However, over the years some of the penetrations have become susceptible to cracking, allowing water and fission products to leak. The response has once again been to replace the component, and reactor operators have now replaced the vessel head in several PWRs. (In recent years corresponding corrosion patterns have been found on penetrations in the bottom of a BWR pressure vessel at Peach Bottom. At that plant boronated water was found to have leaked from the vessel.)

4.4 Instrumentation and control

The instrumentation and control (I&C) technologies in a nuclear plant (referred to in some cases as C&I) are:

- the instruments that interact with the processes in the plant
- the cables carrying the signals from the instruments
- the signal conditioning equipment which modifies the signals into forms useful to the communication channels
- the architecture supporting the transport of signals and data within the plant
- the control room
- the man–machine interfaces
- the procedures
- the control equipment
- the control algorithms
- the computer software used in the monitoring, control, safety, communication and display systems

Safety systems in nuclear power plants require a level of qualification of I&C substantially higher than in monitoring, control, communication and display systems.

Many nuclear power plants were designed and built in the era of analogue I&C equipment, and reactors built in the 1970s, 1980s and even 1990s had very little computer power by modern standards.

In a study for the US industry James D White, of the Oak Ridge National Laboratory, noted that new I&C technologies for nuclear power plants were of considerable interest to the nuclear industry throughout the world. This interest derives from two considerations. The first is that the I&C systems are the windows into the status of the nuclear plant. Since the Three Mile Island accident, the industry had been trying to improve the ability of the operators to grasp the safety status of the plant, particularly during operational upsets. The advent of computer-based monitoring and display systems provided opportunities for advancements that should improve the ability of the operators to understand the plant status and, therefore, improve the operator's ability to make the best decisions during the plant transients, which might otherwise become accidents.

The second consideration is that the nuclear industry is being driven towards computer-based instrumentation and control systems. The driving forces are: decreases in reliability of ageing analogue-based I&C; lack of spare parts because the suppliers have moved on to digital hardware; the promise of higher reliability of digital technologies; and the lure of expanded capabilities of software-based systems.

Other industries preceded the nuclear industry in the use of computer-based I&C. The possible consequences of failure of safety systems in nuclear power plants resulted in a great deal of conservatism in the nuclear industry. Although this conservatism affects the design and regulation of nuclear safety systems, it also influences the design of I&C systems for nuclear power plants. The nuclear industry moves very slowly to make changes in designs.

When the demand for nuclear I&C components declined in the 1980s, many vendors eliminated their nuclear quality assurance programmes and product lines. At the same time, the power industry in general began a transition from analogue to digital technology. As a result, most of today's commercial I&C components rely on digital microprocessors. These microprocessors use embedded software to perform many of the mechanical–electrical functions of traditional analogue hardware. Ageing analogue devices often cannot be repaired or replaced due to lack of parts and documentation.

But while modern digital I&C systems offer superior performance and reliability, digital equipment also tends to become obsolete more quickly than its analogue predecessors, adding the need for qualifying state-of-the-art digital systems to replace early digital devices.

Before the advent of digital I&C, nuclear power plant control rooms had wall panels of dials, gauges, strip chart recorders, alarm lights and switches. The operators moved about the control room: standing to make control changes in the plant, having to walk from panel to panel to read strip chart recorders and to turn switches.

During operational upsets, hundreds of alarms and lights alert the operators about certain limits being exceeded. Alarms are a good example of how operator actions were impeded by their equipment.

- During any significant transient, there were hundreds of alarms sounding and alarm lights lit in the first few minutes. Important indications of abnormal conditions were masked by many less important alarms. This reduced the ability of the operator to locate the most relevant alarms quickly.
- Alarms were frequently caused by the action of the operator, making it difficult to understand which alarms were due to an important initiating event and which were due to operator action.
- Some alarms were due to out-of-service components undergoing maintenance, rather than an unsafe operating condition.
- Alarms were generally not received in a predictable order during fault conditions. The first alarm seen by the operator may not have been the original fault, but

only a secondary consequence of some event. To prevent spurious alarms, the tolerance bands are relatively broad. The initiating event may be under way for some time before the alarm is activated.

A great deal of training is necessary for the operators to be able to discern what has happened and what should happen next. For example, even though they had substantial training, the Three Mile Island operators could not determine the nature of the accident at their plant and made mistakes responding to the situation.

The advent of inexpensive, powerful computers with high-resolution monitors allowed designers to consider control room concepts in which the wall panels are replaced by computers. Computer-based workstations surround the operators in such a manner that they do not have to move from their seats to monitor and control any of the plant's major systems.

Because there is such a large quantity of information that the operator might need, there is a concern that the operator might lose the big picture while searching through the instruments and computer-based displays surrounding him in a cockpit-type control room. To avoid this, most new control room designs include a large diagram of the plant on one of the control room walls to present to all observers the status of the plant's major systems and alarms.

Digital software helps prioritise alarm signals. It also allows operators to move on screen from an overview of the entire plant operation, to an increased level of detail on particular systems. The software may be linked to maintenance management software so the operator can have online information about whether systems flagged up by alarms are undergoing maintenance or could be affected by work elsewhere. Finally, the software can link operators directly with online and prioritised versions of the plant operating manuals, so it is simple for operators to consider their course of action.

In many plants, digital I&C is also backed up by a 'simulator' of the control room where operators can receive regular training on incident or accident scenarios.

Many plant operators now assume that plant I&C systems may be replaced on a 15–20 year basis to deal with obsolescence and take advantage of new capabilities of the systems.

4.5 Life limits

The operating lifetime of a reactor is very different from the way it was planned when the existing reactors were built. Huge or fundamental components like the steam generators or the I&C systems are now no longer seen as fixed for the life of the plant but as potential candidates for upgrade or replacement, if the economics justify it. Some have speculated that BWR internals will be replaced in the near future, or even that the reactor pressure vessel of a PWR is a candidate for replacement – especially in the smaller VVERs, where the vessel is vulnerable to embrittlement of the major welds as the vessel is irradiated over time.

If so much work can be done to extend it, what is the effective lifetime of a reactor?

Many PWR operators are now hoping to keep reactors in operation for up to 60 years, using careful life management programmes, and new reactor designs plan in lifetimes of at least this length.

4.6 Life extension

In the UK, nuclear plant licences are not issued on a time-limited basis. UK reactors undergo an extensive safety inspection once per decade and on this basis receive a continuing licence to operate. British Energy recently extended its planned operation at Dungeness B by 10 years and hopes to extend the operating lives of other AGRs.

In the US and some other countries, however, the licensing regime is somewhat different. The operating licence must be renewed – effectively, the licensing process undergone when the reactor first started up – in order to extend the reactor life and once again the licence is time-limited. Operating licences expire after 40 years but may be extended with the approval of the Nuclear Regulatory Commission. Licence renewals add 20 years. The NRC has indicated that 'substantially all' existing reactors intend to renew their licences. The renewal process has been less burdensome than was once anticipated and is at best only an indicator of the future of particular plants.

Tom Weir, Framatome ANP senior vice president for engineering, told NEI that although many utilities are getting extensions to 60 years, that does not have to be the limit. 'Maximum plant life is unknown and the industry will operate the units as long as they remain safe and economically viable,' Weir said. 'The industry has demonstrated that an additional 20 years is very viable.' He pointed out that when he started in the nuclear business, no one would have even considered steam generator replacements – now they are done routinely. Both technology and construction techniques have changed significantly.

In the meantime, rising electricity prices have meant that some nuclear stations that had been mothballed for many years are coming back into operation. TVA has announced that by 2007 it hopes to bring Browns Ferry 1 back into operation. That reactor has been closed since 1985. In Canada, meanwhile, two units at the Bruce reactor and one at Pickering have already been restarted. In addition, work has begun on refurbishing two other units at Bruce that for many years were thought too expensive to revive because of the extensive maintenance and repair that would be required.

4.7 Uprating

Another recent trend that will result in increased nuclear capacity to help sustain the nuclear share of electricity generated is referred to as capacity uprates. Uprating capacity has been an ongoing process since the inception of the nuclear power industry. Uprates have also occurred in other power sub-sectors such as coal and hydropower.

Present technologies permit uprates of existing nuclear reactors of between 5 per cent and 20 per cent, as has been borne out by upgrades in Sweden and Finland. In the US, projections from the Department of Energy's Energy Information Administration place the near term potential around 4 GWe, based primarily on utility and regulatory announcements. Others such as the Nuclear Energy Institute go as high as 10 GWe. One restriction on higher numbers will be the balance-of-plant considerations and occasionally the economics of the increase. When uprates are viable they provide low-cost increases in plant capacity with little change in operations and maintenance costs.

A summary in Nuclear Engineering International notes that nuclear power producers in the USA have squeezed the equivalent of four new nuclear plants – and hope for at least two to three more plant-equivalents – from their fleet of 103 operating plants with a series of instrumentation and equipment changes that allow additional power production.

The US Nuclear Regulatory Commission (NRC) has approved more than 100 nuclear power plant uprates under its operating reactor regulations. The uprates can be divided into three categories.

- Measurement uncertainty recapture uprates add less than 2 per cent to the plant rating. They are achieved by implementing enhanced techniques for calculating reactor power by using state-of-the-art feedwater flow measurement devices to reduce the degree of uncertainty associated with these measurements. This, in turn, provides for more accurate power calculations.
- Stretch power uprates typically add power increases of 2–7 per cent and stay within the plant's existing design capacity. The actual percentage power increase is plant-specific and depends on operating margins included in the plant design. Stretch uprates usually involve instrumentation setpoint changes, but do not involve major plant modifications.
- Extended power uprates add power increases of between 7 per cent and 20 per cent, and require significant modification to major balance-of-plant equipment such as the high-pressure turbines, condensate pumps and motors, main generators and/or transformers.

Each level of uprate not only produces a higher percentage power increase but also becomes more expensive. Measurement uncertainty recapture generally costs around $1 million, but an extended uprate that goes for the full 20 per cent can cost in excess of $100 million, depending on the equipment that must be modified.

From a physical standpoint, potential uprates depend on what can be put into the core, the condition of the steam generators, and the condition and replacement cost of secondary-side equipment. Each plant must be examined individually to determine how much excess capacity it has, and what equipment must be replaced to tap that capacity. But changing one piece of equipment can require other equipment replacements to assure compatibility.

Longer plant life also makes uprates more attractive financially. Uprate costs range from about $750 to $900 per installed kilowatt, compared with new plant costs of $1,500 to $1,800 per installed kilowatt. These costs become very attractive economically if they can be capitalised over an additional operating period of 20 years or more.

Several major nuclear vendors, including Bechtel, Framatome ANP, GE Energy and Westinghouse, have developed business units that specialise in plant-specific feasibility studies and analyses of uprate potential. These studies typically look at a range of plant parameters, such as fuel design and improved fuel options, as well as secondary side improvements such as steam heaters and turbine upgrades. Some vendors offer full-service contract management through completion of the uprate. Many utilities launch uprate feasibility studies when they also are looking at major plant upgrades such as turbine replacement.

The NRC has developed a standardised process to evaluate licensee submittals. The regulator seeks about five years' advance notice of a potential application, particularly for a stretch or extended uprate application, so the agency can plan its own budget requirements.

So far, approved uprates on all US nuclear plants total about 4,200 MWe, or the equivalent of four new 1,000 MWe plants. Utilities are at various stages of regulatory application for another 1,000 MWe. Nuclear industry observers predict that a further 1,000–2,000 MWe are possible given the present state of technology.

The NRC has approved 105 uprate applications since 1977. The first extended uprate, a 6.3 per cent power increase for the Monticello plant, was approved in 1998. The NRC approved 17.8 per cent increases for Quad Cities units 1 and 2 in 2001, and the first 20 per cent increase – for the Clinton plant – in 2002.

So far, the majority of the uprates have been on BWRs, which had more excess capacity margin to recover. However, the NRC has seen more activity on PWRs in recent years.

Progress Energy completed a four-year power uprate at Brunswick 1 and 2 in Southport, North Carolina, adding a total of 232 MWe to the output of the two-unit BWR plant. Power uprate and reliability improvements during the recently completed unit 2 outage included replacing three main power transformers and other switchyard improvements. Progress also upgraded condensate and feedwater systems to support uprated power conditions.

Chapter 5
Fuelling the reactor

5.1 Nuclear fuel

Coal is a mixture of organic substances and each deposit has its own characteristics of carbon and water content, along with widely varying proportions of other substances such as sulphur. It undergoes little processing until it reaches the power plant, where it will be ground and prepared for feeding into the plant. The operator must balance coal from different sources with different characteristics, to get the best possible burning qualities, with the most acceptable mix of residues in the resulting smoke and ash.

Uranium fuel, in contrast, is made from a chemically simple metal salt denoted UO_2 (meaning that there are two atoms of oxygen for each atom of uranium), contained in a complex metal framework. All the processing to prepare the uranium is done ahead of its arrival at the plant. The fuel preparation is a complex and sometimes energy-intensive process.

5.2 Mining uranium

Uranium is not a scarce resource. There are many deposits that can be mined economically, even at the very low uranium prices that were offered in the 1990s, and extensive reserves available at higher cost.

As with other mineral resources, 'known' reserves generally increase in a cycle of several decades, moving in step with the market for the mineral. As current sources are exploited and the remaining well-characterised sources become progressively more expensive to mine, the market price of the product increases. The increasing price prompts mining and exploration companies to seek out new sources of the mineral.

At present, the largest economically recoverable sources of uranium are in Australia, which has around 30 per cent of the world's supply, with Kazakhstan having around 17 per cent and Canada around 12 per cent. Somewhat smaller deposits are found in South Africa (8 per cent), Namibia (6 per cent), Brazil (4 per cent), Russia (4 per cent), the USA (3 per cent) and Uzbekistan (3 per cent). Uranium is also mined and milled in small amounts in many other countries.

Canada is the largest producer of uranium, although it has fewer reserves than Australia. Canada has almost completed a transition from second-generation uranium mines (started 1975–83) to new high-grade ones, all in northern Saskatchewan. The Saskatchewan government actively encourages and supports uranium mining in the province, where it is found to be environmentally acceptable. This reversed a previous policy of the New Democratic Party in the early 1990s to phase out uranium mining.

Within Canada, Cameco operates the McArthur River mine, which started production at the end of 1999. Its ore is milled at Key Lake, which once contributed to 15 per cent of world uranium production but has now been mined out. Its other former mainstay is Rabbit Lake, which still has some reserves at Eagle Point, where mining resumed in mid-2002 after a three-year break.

Cogema Resources operates the McClean Lake mine, which started production in mid-1999. Its Cluff Lake mine has now closed and is being decommissioned.

There are further new uranium projects coming into production in the next few years in Northern Saskatchewan that could increase Canada's share of world uranium production to nearly half.

Figure 5.1 McClean Lake mine site. Saskatchewan, Canada [AREVA Archives/DR]

Cigar Lake will be a 450 m deep underground mine in poor ground conditions, using ground freezing and high-pressure water jets for excavation of ore. The full construction licence was issued in December 2004 and construction began in early 2005. Cogema's Midwest mine was initially planned to be underground, utilising ground freezing and water jet boring but may now be mined as an open pit. Finally, a small deposit at Dawn Lake is further from development. Grades up to 30 per cent U_3O_8 at depths of 280 m have been reported nearby.

Australia, with the largest uranium resource, began mining ores in the 1930s to recover minute amounts of radium for medical purposes. Uranium ores as such were mined and treated in Australia from the 1950s until 1971. Radium Hill, SA, Rum Jungle, NT and Mary Kathleen, Queensland were the largest producers of uranium.

The development of civil nuclear power stimulated a second wave of exploration activity in the late 1960s. New contracts for uranium sales (to be used for electric power generation) were made by Mary Kathleen Uranium Ltd, Queensland Mines Ltd and Ranger Uranium Mines Pty Ltd, in the years 1970–72. The Commonwealth Government announced in 1977 that new uranium mining was to proceed, commencing with the Ranger project in the Northern Territory. This mine opened in 1981.

Following the 1983 federal election, the Australian Labor Party (ALP) won power and in 1984 the ALP National Conference amended the Party platform to what became known as 'the three mines policy', nominating Ranger, Nabarlek and Olympic Dam as the only projects from which exports would be permitted. Provisional approvals for marketing from other prospective uranium mines were cancelled. This policy persisted until the Liberal–National Party Coalition government came to power in 1996.

During 1988 the Olympic Dam project, then a joint venture of Western Mining Corporation and BP Minerals, commenced operations. This is a large underground mine in central South Australia, producing copper, gold and uranium, owned by BHP Billiton, following its 2005 takeover of WMC Resources.

Following the 1996 change in government policy, three other projects were brought forward: Jabiluka, in the Northern Territories, and two mines, Honeymoon and Beverley, in South Australia. Jabiluka will be an extension of the Ranger operation but awaits Aboriginal approval for development. The last two are small *in situ* leach operations.

5.2.1 Conventional mining

The mined rocks pass through several stages for the extraction of uranium.

Sweetwater Mill in Sweetwater County, Wyoming, USA is managed and operated by Kennecott Uranium Company, according to the Wyoming Mining Association. It is the only remaining conventional uranium mill in Wyoming – one of six conventional uranium mills in the USA – and it has a typical milling process.

The other uranium mills (and their owners) in the USA are the Ambrosia Lake Mill near Grants, New Mexico (Rio Algom Mining Corporation/now in reclamation),

Ford Mill near Ford, Washington (Dawn Mining Company), Canon City Mill near Canon City, Colorado (Cotter Corporation), White Mesa Mill near Blanding, Utah (International Uranium (USA) Corp.) and the Shootaring Canyon (Tickaboo) Mill near Tickaboo, Utah (US Energy Corporation).

First the rock is ground, using a grinding mill, a crusher, rod mills and ball mills. Ground ore leaves the mill mixed with water at a design density of 70 per cent solids and a maximum particle size of 1 cm. The maximum discharge size is determined by the size of the mill discharge grates.

The ground rock is then treated in a leach circuit to dissolve the uranium out of the crushed rock. At the Sweetwater Mill the slurry is treated in rubber-lined steel leach tanks with steam, sodium chlorate and sulphuric acid to dissolve the uranium. The dissolving of the uranium is a slow process and the design of the 10 leach tanks is such that the slurry will be agitated in the tanks for 12–16 h before being pumped to another circuit. Each of the leaching tanks is rubber lined and measures 22 feet 9 inches in diameter and is 22 feet high.

Not all uranium mills use rubber-lined steel leach tanks. The Shootaring Canyon (Tickaboo) Mill, for example, uses wood stave leach tanks.

In the next process (known as the decantation circuit at Sweetwater Mill), the slurry is passed through a series of tanks in which the liquid containing the dissolved uranium is separated from the remaining solids from the rock. This is accomplished with de-aerators, rake mechanisms, overflow launders, underflow and overflow pumps, and the flocculant addition and dilution system. The uranium-containing liquid (known as the overflow) is clarified and sent to the solvent extraction portion of the circuit. The slurry, now minus the uranium, is pumped to the tailings impoundment.

At Sweetwater the solvent extraction circuit is housed in a separate building because the organic phase consists largely of kerosene, which is flammable. The overflow from counter-current decantation is clarified and mixed with a liquid mixture consisting of kerosene, isodecanol and amine. It is essential that the aqueous overflow sent to the solvent extraction circuit be clean and thoroughly filtered. Suspended solids in the aqueous overflow can cause fouling problems in the solvent extraction circuit.

The uranium leaves the aqueous phase and transfers to the organic phase in a process called liquid ion exchange. In some mills, this process is accomplished through the use of ion exchange columns and solid ion exchange resins. The uranium is concentrated in this step. It is then further concentrated by transferring it from the organic phase to a second aqueous phase. The uranium is then precipitated from the aqueous phase with ammonia.

This yields a so-called yellowcake slurry. Yellowcake is a form of uranium oxide denoted U_3O_8 (i.e. with three parts uranium to eight parts oxygen) and it is a granular solid. The yellowcake is dried, barrelled and shipped to the converter.

5.2.1.1 Mining waste

The wastes arising from uranium mining, according to the World Nuclear Association, are in most respects the same as in mining any other metalliferous ore. From open cut

Figure 5.2 *Yellowcake on a belt filter at the ore processing plant operated by Société des Mines de Jouac (SMJ). Haute-Vienne Department, France [AREVA Archives/Philippe Lesage]*

mining, there are substantial volumes of barren rock and overburden waste. These are placed near the pit and either used in rehabilitation or shaped and revegetated where they are.

Solid waste products from the milling operation are known as tailings. They comprise most of the original ore and they contain most of the waste's radioactivity. In particular they contain all the radium present in the original ore. In an underground mine they may be first cycloned to separate the coarse fraction which is used for underground fill. The balance is pumped as a slurry to a tailings dam, which may be a worked-out pit as at Ranger and McClean Lake.

When radium undergoes natural radioactive decay one of the products is radon gas. Because radon and its decay products (daughters) are radioactive and because the tailings are now on the surface, measures are taken to minimise the emission of radon gas. During the operational life of a mine the material in the tailings dam is usually covered by water to reduce surface radioactivity and radon emission (though with lower-grade ores neither pose a hazard at these levels).

On completion of the mining operation, it is normal for the tailings dam to be covered with some 2 m of clay and topsoil to reduce radiation levels to near those normally experienced in the region of the orebody and for a vegetation cover to be established. At Ranger and Jabiluka in North Australia, tailings will be returned underground, as was done at the now-rehabilitated Nabarlek mine. In Canada, ore treatment is often remote from the mine that the new ore comes from, and tailings are emplaced in mined out pits wherever possible, and engineered dams otherwise.

Run-off from the mine stockpiles and waste liquors from the milling operation are collected in secure retention ponds for isolation and recovery of any heavy metals or other contaminants. The liquid portion is disposed of either by natural evaporation or recirculation to the milling operation. Process water discharged from the mill contains traces of radium and some other metals which would be undesirable in biological systems downstream. This water is evaporated and the contained metals are retained in secure storage. During the operational phase, such water may be used to cover the tailings while they are accumulating.

5.2.2 In situ *leaching*

In recent years, a chemical method of mining uranium, known as '*in situ* leaching', has made it possible to exploit deposits that may not otherwise have been economical to mine, as conventional techniques require that the mine be situated close to a uranium mill.

In situ mining extracts uranium from porous sandstone aquifers by reversing the chemical processes which deposited the uranium.

To be mined *in situ*, the uranium deposit must occur in permeable sandstone aquifers. These sandstone aquifers provide the 'plumbing system' for both the original emplacement and the recovery of the uranium. The uranium was emplaced by weakly oxidising ground water which moved through the geologic formation. To effectively extract uranium deposited from ground water, a company must first thoroughly define this system and then design well fields that best fit the natural hydrogeological conditions.

Once the geometry of the ore bodies is known, the locations of injection and recovery wells are planned to ensure that as much of the uranium as possible is in contact with the well fields.

Following the installation of the well field, a leaching solution (or lixiviant), consisting of native ground water containing dissolved oxygen and carbon dioxide, is delivered to the uranium-bearing strata through the injection wells. Once in contact with the mineralisation, the lixiviant oxidises the uranium minerals, which allows the uranium to dissolve in the ground water. Production wells, located between the injection wells, intercept the lixiviant with its load of uranium and pump it to the surface. A centralised ion-exchange facility extracts the uranium from the lixiviant.

The barren lixiviant, stripped of uranium, is regenerated with oxygen and carbon dioxide and recirculated for continued leaching. The ion exchange resin, which becomes 'loaded' with uranium, is stripped or eluted. Once eluted, the ion exchange resin is returned to the well field facility.

During the mining process, slightly more water is produced from the ore-bearing formation than is reinjected. This net withdrawal, or 'bleed', produces a cone of depression in the mining area, controlling fluid flow and confining it to the mining zone. The mined aquifer is surrounded, both laterally and above and below, by monitor wells which are frequently sampled to ensure that all mining fluids are retained within the mining zone. The 'bleed' also provides a chemical bleed on the aquifer to limit the buildup of species such as sulphate and chloride which are affected by the leaching process.

The 'bleed' water is treated for removal of uranium and radium. This treated water is then disposed of through waste water land application, or irrigation. A very small volume of radioactive sludge results; this sludge is disposed of at an NRC licensed uranium tailings facility.

The ion exchange resin is stripped of its uranium, and the resulting rich eluate is precipitated to produce a yellowcake slurry. This slurry is dewatered and dried to a uranium concentrate which is then placed in a drum.

At the conclusion of the leaching process in a well field area, the same injection and production wells and surface facilities are used for restoration of the affected ground water. Ground water restoration is accomplished in three ways. First, the water in the leach zone is removed by 'ground water sweep', and native ground water flows in to replace the removed contaminated water. The water that is removed is again treated to remove radionuclides and disposed of in irrigation.

Second, the water that is removed is processed to purify it, typically with reverse osmosis, and the pure water is injected into the affected aquifer. Third, the soluble metal ions which resulted from the oxidation of the ore zone are chemically immobilised by injecting a reducing chemical into the ore zone, immobilising these constituents *in situ*. Ground water restoration is continued until the affected water is suitable for its pre-mining use.

Once mining is complete, the aquifer is restored by pumping fresh water through the aquifer until the ground water meets the pre-mining use.

5.3 Conversion

After the yellowcake is produced, the next step is conversion into uranium hexafluoride (UF_6). Uranium hexafluoride is a chemical compound consisting of one atom of uranium combined with six atoms of fluorine. It is the chemical form of uranium that is used during the uranium enrichment process. This is because UF_6 is the only uranium compound that exists as a gas at a suitable temperature. Within a reasonable range of temperature and pressure, it can be a solid, liquid or gas. Solid UF_6 is a white, dense, crystalline material that resembles rock salt.

Impurities are removed and the resulting uranium trioxide (UO_3) is converted to the intermediate compound uranium tetrafluoride (UF_4) via a series of process steps. The UF_4 is then combined with fluorine to create the UF_6 product. The UF_6 is then pressurised and cooled to a liquid. In its liquid state it is drained, for example, into 14-ton cylinders where it solidifies after cooling for approximately five days. The cylinder, with UF_6 in its solid form, is then shipped to an enrichment plant.

Conversion plants operate in the USA, Brazil, China, France (Pierrelatte), UK (Springfields), South Africa, Japan, Canada and Russia.

As with mining and milling, the primary risks associated with conversion are chemical and radiological. Strong acids and alkalis are used in the conversion process,

Figure 5.3 Fluorine electrolysis cells at Comurhex-Pierrelatte plant for conversion of UF_4 into UF_6. Tricastin, France [AREVA Archives/DR]

which involves converting the yellowcake (uranium oxide) powder to very soluble forms, leading to possible inhalation of uranium. In addition, conversion produces extremely corrosive chemicals that could cause fire and explosion hazards.

5.4 Enrichment

Once converted to the UF_6 form that can be easily turned into a gas, the uranium can be processed to increase the proportion of ^{235}U to a suitable level of 'enrichment' for use in fuel. Two processes are used – gaseous diffusion and centrifuges – and both take advantage of the tiny difference in mass between ^{235}U and ^{238}U.

5.5 Gaseous diffusion

In the gaseous diffusion enrichment plant, the solid uranium hexafluoride (UF_6) from the conversion process is heated in its container until it becomes a liquid. The container becomes pressurised as the solid melts and UF_6 gas fills the top of the container. The UF_6 gas is slowly fed into the plant's pipelines where it is pumped through special filters, called barriers or porous membranes. The holes in the barriers are so small that there is barely enough room for the UF_6 gas molecules to pass through. The isotope enrichment occurs because the lighter UF_6 gas molecules (with the ^{234}U and ^{235}U

atoms) tend to diffuse faster through the barriers than the heavier UF_6 gas molecules containing ^{238}U.

It takes many hundreds of barriers, one after the other, before the UF_6 gas contains enough uranium-235 to be used in reactors. At the end of the process, the enriched UF_6 gas is withdrawn from the pipelines and condensed back into a liquid that is poured into containers. The UF_6 is then allowed to cool and solidify before it is transported to fuel fabrication facilities where it is turned into fuel assemblies for nuclear power reactors.

The only gaseous diffusion plant in operation in the United States is in Paducah, Kentucky. A similar plant was in operation in Piketon, Ohio, but it was shut down in March 2001. Both plants are leased by the United States Enrichment Corporation (USEC) from the Department of Energy and have been regulated by the NRC since 4 March 1997. In France, Eurodif operates a gas diffusion plant.

5.6 Gas centrifuge

The gas centrifuge uranium enrichment process uses a large number of rotating cylinders in series and parallel formations. Centrifuge machines are interconnected to form trains and cascades.

In this process, UF_6 gas is placed in a cylinder and rotated at a high speed. This rotation creates a strong centrifugal force so that the heavier gas molecules (containing ^{238}U) move towards the outside of the cylinder and the lighter gas molecules (containing ^{235}U) collect closer to the centre. The stream that is slightly enriched in ^{235}U is withdrawn and fed into the next, higher stage, while the slightly depleted stream is recycled back into the next, lower stage. Significantly more ^{235}U enrichment can be obtained from a single unit gas centrifuge than from a single unit gaseous diffusion stage.

The European enrichment company Urenco has pioneered the use of the gas centrifuge and the technology is both more compact and much more energy efficient than the gaseous diffusion technology.

The US is in the process of phasing out its gas diffusion plants and both Louisiana Energy Services (LES) and USEC Inc have submitted licence applications for centrifuge plants. USEC Inc was granted a licence in February 2004 for a demonstration and test gas centrifuge plant, which is currently under construction. Russia also uses its own centrifuge enrichment technology.

Once enriched, the UF_6 is transported to a fuel fabrication plant where it is converted to a different uranium oxide – UO_2 – which will be used in the reactor fuel.

5.7 Fuel fabrication

Fuel fabrication processes for the most common reactor types, pressurised- and boiling-water reactors, are similar. Enriched UF_6 is transported to a fuel fabrication plant where it is converted to uranium dioxide (UO_2) powder. This uranium compound is used in reactors because it is stable to extremely high temperatures of many hundred degrees. Higher temperatures cause changes in the oxide's granular structure but it is chemically stable to over 2,000 °C.

The oxide powder is cold pressed at 150–300 MPa to form pellets of 50–60 per cent theoretical density. These so-called green pellets are then sintered in a hydrogen–argon atmosphere at 1,600–1,700 °C for 5–10 h to produce pellets of 95–96 per cent density. They finally undergo a further annealing in hydrogen gas. Although designs vary from core to core, typical final dimensions of fuel pellets are in the order of 1 cm in diameter and 1 cm in height.

5.8 Cladding

Fission of the fissile species in the fuel results in the production of radioactive fission products. Solid fission products are easily retained in the fuel matrix, but gaseous fission products may diffuse from the fuel and be released into the coolant system. This represents a significant hazard to the general public.

Particularly important are radioactive isotopes of iodine which, as airborne contaminates, are readily absorbed in the body and may result in a significant dose to internal organs. It is therefore necessary to prevent the release of these fission products from the fuel. This is accomplished by sealing the fuel in zircaloy cladding.

Zircaloy is an alloy of zirconium and is used in two principal forms, Zircaloy-2 and Zircaloy-4. Zircaloy-2 consists of a zirconium sponge with 1.5 wt.% tin, 0.12 wt.% iron, 0.01 wt.% chromium and 0.05 wt.% nickel. Zircaloy-4 is similar to Zircaloy-2 but without the addition of nickel and has an iron content of 0.18 wt.%. Zirconium has the advantages of having high thermal conductivity, low neutron absorption cross-sections, good multiaxial rupture strength, good creep strength and ductility. The addition of alloying elements increases the corrosion resistance of the clad in high-temperature aqueous environments over that of pure zirconium.

The difference between Zircaloy-2 and Zircaloy-4 is primarily due to nickel, which tends to absorb hydrogen so that Zircaloy-4 absorbs less hydrogen than Zircaloy-2 during high-temperature water corrosion. Zircaloy-4 is used in pressurised light-water reactor applications and Zircaloy-2 is used in boiling-water reactors.

5.9 Fuel pin construction

The basic fuel pin arrangement consists of a zircaloy tube capped at both ends, with sintered UO_2 pellets stacked inside. An initial (unirradiated) gap is designed between the fuel and the clad wall to accommodate swelling of the fuel due to the build-up of fission products and differential thermal expansion between the fuel and the clad. The top of the fuel pin is void of fuel to create a plenum for the retention of fission product gases throughout the life of the pin, and a spring is provided at the top of the pellet stack to minimise motion of the pellets, particularly during the shipping of new fuel. The rod is internally pressurised to 10 atm of He to minimise the differential pressure experienced across the clad during operation.

Figure 5.4 MOX fuel-rod assembly area at Melox fuel fabrication plant. Bagnols-sur-Ceze, France [AREVA Archives/Sidney Jazequel]

In a CANDU reactor, fuel pellets are loaded into 28 or 37 half-metre long rods grouped into a cylindrical fuel bundle. Twelve bundles lie end to end in a fuel channel in the reactor core. A Bruce 790 MWe CANDU reactor contains 480 fuel channels composed of 5,760 fuel bundles and over 5 million fuel pellets.

Fuel fabrication processes for other reactor designs – Magnox and AGR, for example – are somewhat different, and the differing requirements of fast reactors and new reactor designs are also likely to require changes in fuel design and fabrication.

5.10 Future sources of uranium

As was noted at the start of the book, uranium is not a rare element. Assessing how much is available for use as nuclear fuel, and whether it is economically viable, is more problematic.

Some figures are fairly well accepted. The IAEA says ‘known uranium reserves with reactors operating primarily on a once-through cycle without reprocessing of spent fuel assure a sufficient fuel supply for at least 50 years at current levels of use, the same order of magnitude as today’s proven reserves of natural gas and oil. Estimates of additional undiscovered (speculative) reserves could add more than 100 years.’

The main source for information on the resource, the OECD-NEA’s so-called Red Book, lists ‘known resources’ by recovery cost. It says there are around 3.6 million tonnes of uranium recoverable at less than $80 per kgU and another million tonnes recoverable at $80–130/kgU. There are around 400 reactors in operation worldwide and they require around 76,000 kgU per year – which suggests there is fuel available for between 40 and 50 years.

The Red Book’s 4.6 million tonnes refer to deposits that are well known, well characterised and well costed – referred to as either reasonably assured, or inferred, resources – and in fact in most cases those resources are already being mined.

Further to this, the OECD-NEA also lists resources that are hypothetical or speculative. These add an estimated 9.8 million tonnes to the reserves. These speculative reserves would therefore provide enough fuel so that double the current number of reactors could be brought on line (around 800 new reactors worldwide) and each of those could have a lifetime of 60 years.

This estimate of deposits is speculative – and in fact estimates have shrunk slightly in the last two years – but it is not comprehensive. Australia, for example, which has some of the largest uranium resources, does not report its uncharacterised reserves to the NEA for political reasons.

Over and above these deposits, there are also what the Red Book describes as ‘unconventional’ resources. This includes phosphate deposits, and the Red Book estimates that uranium recoverable as by-product from phosphates is around 22 million tU. Even in a greatly expanded nuclear industry this could clearly provide enough fuel for centuries. Technology for extracting uranium from phosphates has been demonstrated on an industrial scale in Belgium and in the USA and the process is considered to be mature. The cost of uranium extraction in a plant producing 100 tU per year is put at $60–100 per kgU.

The largest potential source is seawater, whose uranium resource is generally estimated at around 4,000 million tU. The cost of extracting uranium from seawater is higher than that of phosphate reserves. Julian Steyn of Energy Resources International put the figure at $600–1,200 per kgU, but others place it at $300 per kgU.

The extraction costs of uranium from unconventional sources look extremely high when uranium prices are running at around $80 per kg. But it must be recalled that the uranium price is currently a very small part of the fuel cost, far outweighed by the costs of enrichment, and that fuel price also includes conversion, fuel manufacture and all the steps detailed above. What is more, fuel is a very small proportion of the price of nuclear electricity – perhaps a tenth. So the uranium cost could increase several hundred per cent before it added significantly to the price of nuclear electricity.

These figures take no account of the more efficient use of uranium in recent reactor designs. They also take no account of current or future developments in mining. Since the early 1990s there had been little uranium exploration or mine development that could characterise new uranium sources or reduce mining costs – in fact, a significant proportion of the uranium has been sourced from reductions in US and Russian defence programmes. The effect of this addition has been to almost halt investment in uranium mining, and as with other resources, like coal or oil, additional deposits are usually found in periods when the resource is scarce, when high resource prices prompt mining companies to prospect.

Julian Steyn of Energy Resources International in the USA is an expert on the world uranium market. He says: 'If we mount programmes we should be able to increase the reserve base but the lead time will be 10–20 years and – not unlike the oil and gas industry – it depends on how much money is being invested.' The next few years may see new resource discoveries, simply because it is likely that a market will be there for them. Julian Steyn says that 'Now – since there is talk of a nuclear renaissance – there is a gold rush fever. There were 100 claims in the western USA over the last 100 years. This year in the western USA there have been 2,000 claims already. And people are dusting off old shutdown mines.' Uranium prices have jumped from $7 per lb in 2000 to $33 per lb.

Higher extraction costs are also likely to mean that there will be developments in mining techniques. The only major change in the recent years of depressed market has been the development of '*in situ* leaching', instead of standard extraction. It is a slightly less efficient method for large deposits but can be used to extract uranium in thin seam deposits. The technique is being used, for example, in Kazakhstan, where 'there is a huge quantity available for *in situ* leaching. One mine is already being pursued', says Julian Steyn.

5.10.1 Assessing the estimates

Because uranium has not been sought in the past few years and the known reserves have changed little, there is still debate outside and inside the industry about how much is economically available. A study by the Massachusetts Institute of Technology, for example, recently concluded that 'the world-wide supply of uranium ore is sufficient to fuel the deployment of 1,000 reactors over the next half century and to maintain this level of deployment over the 40-year lifetime of this fleet.' But at the same time, the US government's 'Generation IV' reactor development programme shows requirements of uranium overtaking 'known resources' by the middle of the next century, even with very little growth in the number of reactors.

The nuclear industry is divided over uranium's potential because it is divided over the course the industry should take. The MIT study comes out strongly for a once-through or 'open' cycle, as currently employed in the USA, where spent fuel from the reactors is sent direct for disposal. The report says 'The once through cycle has advantages in cost, proliferation, and fuel cycle safety, and is disadvantageous only in respect to long-term waste disposal.' It describes two alternative 'closed cycles', one in which spent fuel is reprocessed and the recovered plutonium and uranium are burnt in conventional reactors, and one in which the reprocessed materials are used to fuel 'fast breeder' reactors that create more plutonium fuel.

The MIT report says 'Closed fuel cycles may have an advantage from the point of view of long-term waste disposal and, if it ever becomes relevant, resource extension. But closed fuel cycles will be more expensive than once-through cycles, until ore resources become very scarce. This is unlikely to happen, even with significant growth in nuclear power, until at least the second half of this century, and probably considerably later still.'

The Generation IV reactor programme, with support from a dozen countries worldwide (including the UK individually and via the EU), is following up six designs, all with some form of closed cycle fuel programme. The MIT report states 'This decision implies a major re-ordering of priorities of the US Department of Energy nuclear R&D programs.' There are also associated programmes on fuel cycles and on using the new designs to generate hydrogen. The MIT report disagrees with this approach and says 'For the next decades, government and industry in the US and elsewhere should give priority to the deployment of the once-through fuel cycle. Expensive programs that plan for the development or deployment of commercial reprocessing based on any existing advanced fuel cycle technologies are simply not justified on the basis of cost, unproven safety, proliferation risk, and the waste properties of a closed cycle compared to the once-through cycle.'

It must also be recalled that the long-standing attitudes of the various countries to the fuel cycle heavily influence decisions on the future route. Countries like Japan, France and the UK with existing infrastructure and expertise in fuel-cycle reprocessing tend to be more favourably inclined towards reprocessing, unlike countries such as the USA that have little infrastructure for spent fuel recycling.

5.10.2 When to decide?

Future fuel cycles are of immediate interest to reactor operators. That is because fuel management in the reactor will be influenced by its ultimate fate. At the moment, most countries foresee a once-through cycle. That means operators want to get as much power as possible out of each fuel rod, so the trend has been towards leaving it in the reactor longer to get increased 'burnup'. It is economically efficient in a once-through cycle and fuel assemblies are designed to support it, so fuel residency in the reactor has been increasing towards the two-year mark.

However, the radioactive products produced in the reactor include uranium-236 and uranium-232, both of which cause problems in re-using the uranium in any 'closed' cycle. The uranium has to be enriched separately from fresh uranium and this is one reason why uranium recovered from existing reprocessing plants is seldom re-used in practice. In the UK the original fuel baseload for the AGR reactors was sourced from recycled uranium and was enriched in the now-redundant gas diffusion enrichment plant at Capenhurst. The economics of using recycled uranium are currently under review due to the rising cost of uranium in world markets.

In France EdF has made provision to store reprocessed uranium (RepU) for up to 250 years as a strategic reserve. Currently, reprocessing of 1,150 tonnes of EdF used fuel per year produces 8.5 tonnes of plutonium (immediately recycled as mixed oxide – MOX – fuel) and 815 tonnes of RepU per year. Of this about 650 tonnes is converted into stable oxide form for storage. EdF has demonstrated the use of RepU in its 900 MWe power plants, but it is currently uneconomic because it has higher conversion costs and enrichment needs to be separate because of ^{232}U and ^{236}U impurities (the former gives rise to gamma radiation, the latter means higher enrichment is required).

Where uranium recovered from reprocessing spent nuclear fuel is to be re-used, it needs to be converted and re-enriched. This is complicated by the presence of impurities and two new isotopes in particular: ^{232}U and ^{236}U, which are formed by neutron capture in the reactor. Both decay much more rapidly than ^{235}U and ^{238}U, and one of the daughter products of ^{232}U emits very strong gamma radiation, which means that shielding is necessary in the plant. ^{236}U is a neutron absorber which impedes the chain reaction and means that a higher level of ^{235}U enrichment is required in the product to compensate. Being lighter, both isotopes tend to concentrate in the enriched (rather than depleted) output, so reprocessed uranium which is re-enriched for fuel must be segregated from enriched fresh uranium.

5.11 Managing spent fuel

When fuel has finished its residency time inside the nuclear reactor it is removed and is eventually sent for reprocessing or final disposal. First, however, it is stored on-site in a water-filled pool several metres deep. The water acts as both coolant and shielding, because the residual heat in the fuel is initially very high as fission products continue to decay. The spent fuel continues to produce heat for years, but the heat production rate decays fairly quickly at first, as short-lived elements complete their decay.

The next stage of spent fuel management depends on the fuel route chosen by the plant operators. In the UK, AGR and Magnox fuel cannot be stored indefinitely: instead, it must be sent for reprocessing.

According to Australia's Uranium Information Centre, for most types of fuel, reprocessing occurs anything from 5 to 25 years after reactor discharge.

All commercial reprocessing plants use the well-proven hydrometallurgical Plutonium Uranium Extraction (PUREX) process. This involves dissolving the fuel elements in concentrated nitric acid. Chemical separation of uranium and plutonium is then undertaken by solvent extraction steps (neptunium can also be recovered if required). Once extracted, the plutonium and uranium can be returned to the input side of the fuel cycle – the uranium to the conversion plant prior to re-enrichment and the plutonium straight to fuel fabrication.

After reprocessing, the recovered uranium may be handled in a normal fuel fabrication plant (after re-enrichment), but the plutonium must be recycled via a dedicated mixed oxide (MOX) fuel fabrication plant. In France the reprocessing output is coordinated with MOX plant input, to avoid building up stocks of plutonium. (If plutonium is stored for some years the level of americium-241, the isotope used in household smoke detectors, will accumulate and make it difficult to handle through a MOX plant due to the elevated levels of gamma radioactivity.)

Figure 5.5 Spent-fuel transportation cask at Cogema-La Hague reprocessing plant. Cherbourg, France [AREVA Archives/Philippe Lesage]

The remaining liquid after the plutonium and uranium are removed is high-level waste, containing about 3 per cent of the used fuel in the form of fission products and minor actinides (neptunium, americium and curium). It is highly radioactive and continues to generate a lot of heat. It is conditioned by calcining and incorporation of the dry material into borosilicate glass, then stored pending disposal. In principle any compact, stable, insoluble solid is satisfactory for disposal.

5.11.1 Developments of PUREX

Another version of PUREX has the minor actinides (americium, neptunium, curium) being separated in a second aqueous stage and then directed to an accelerator-driven system cycling with pyroprocessing for transmutation. The waste stream then contains largely fission products.

Another variation of PUREX is being developed by the US Department of Energy for civil wastes. In this, only uranium is recovered (hence this is known as the UREX or UREX+ process) initially for recycle or for disposal as low-level waste. Iodine and technetium may also be recovered at the head end. The residue is treated to recover plutonium (or plutonium plus neptunium) for recycling in conventional reactors, then the other actinides for transmutation in fast reactors. The fission products then comprise most of the high-level waste. A major goal of this system is to keep the plutonium with other transuranics, which if not utilised in conventional reactors may be destroyed by burning in a fast neutron reactor. The intermediate recycle stage could involve plutonium being recovered for recycling commercially as fuel for normal reactors, as in Europe. At present all this remains contrary to official US policy.

The benefits of reprocessing are that it allows uranium and plutonium to be recovered from the spent fuel to be used as fresh fuel – although opponents argue that this is in practice no benefit, as it is seldom reused but instead its existence presents new dangers. It is also intended to reduce the volume of high-level waste that must be sent for final disposal. However, the by-products of reprocessing also include liquids that are challenging to handle both chemically and radiologically. Such high-level liquid waste is now managed by solidifying it in a glass matrix prior to it being sent to a final repository.

The UK, France, Germany and other European countries initially planned to reprocess all their spent fuel, although that policy has now changed in some countries. In countries that were once part of Comecon initially Russia was the fuel supplier and as part of its supply agreement was expected to take back spent fuel for reprocessing. The fall of the Soviet Union altered that plan and most countries now – by choice or necessity – are following the so-called once-through cycle. In Asia, Japan is in favour of reprocessing spent fuel and after sending it to the UK and France for reprocessing is developing its own reprocessing plant at Rokkasho-mura.

5.11.2 The once-through cycle

The US decided not to pursue reprocessing for managing its spent fuel. Instead, it decided the fuel would remain on-site at the reactor until, in return for a levy on all power sold from nuclear stations, the US government would 'take title' to the fuel and would send it direct to a final disposal site which it would develop.

In fact (see Chapter 6) the US government programme has been delayed and it has resulted in much-extended storage at the reactor sites.

The US lack of a repository has placed nuclear power plants in the position of storing more used fuel than expected for longer than originally intended. The result is that many nuclear plants, each of which produces an average of about 20 tonnes of used fuel annually, are running out of storage capacity. By the end of 2006, about 60 units will have no more storage space in their used fuel pools, and by the end of 2010, 78 will have exhausted their storage capacity.

When a plant's used fuel pool nears its designed capacity, a utility has two options. The first is to re-rack the fuel storage pond. In this process the stored assemblies in the spent fuel pond are moved closer together. More than 130 re-rackings have been done at various nuclear plant sites, but re-racking has its limitations. Strict requirements exclude the possibility of unintended nuclear chain reactions, prevent overheating of the pool and ensure that the pool's structural/earthquake resistance capability is not exceeded. These requirements restrict the extent to which the assemblies can be moved closer together, thus limiting the additional space that can be gained. In some cases operators ensure that fuel in the spent fuel pool stays well within criticality limits by including boron-impregnated neutron absorbers between the fuel racks.

Eventually, used fuel pools reach their capacity. Building a new used fuel pool is not an option. It is too costly and almost impossible to fit a new pool into the plant layout. Although a few companies have shipped used fuel from one plant to another with extra storage capacity, this option obviously is not available to everyone. Most nuclear plants have used the additional pool capacity gained by re-racking, and a growing number have built or are building dry storage facilities on site.

The second option is to remove the fuel from the pool and store it in large rugged containers made of steel or steel-reinforced concrete, 18 or more inches thick. The containers use materials like steel, concrete and lead – instead of water – as a radiation shield. Depending on the design, a dry container can hold from 7 to 56 12-foot-long fuel assemblies.

The US Nuclear Regulatory Commission has approved several designs for use by utilities. The containers have a 20-year licence. After 20 years they must be inspected, and with NRC approval the licence could be extended.

Loaded containers are filled with an inert gas, sealed and stored either on reinforced concrete pads or inside steel-reinforced concrete bunkers. The containers are designed to withstand natural disasters, such as tornadoes, hurricanes and floods, and to prevent the release of radioactivity. The designs require no mechanical devices for cooling and ventilation.

There are, however, some concerns. For example, the use of the site. Unless DOE fulfils its legal obligation to provide used fuel storage and/or disposal, used fuel may have to remain in on-site dry storage facilities beyond the envisioned 20–40 year period. This would prevent some nuclear plant sites from being used for other purposes after the plants are decommissioned. For these reasons, some state governments have opposed the licensing of additional dry storage. The cost of dry storage is another concern. Designing, building and licensing a dry storage facility at a plant site requires an initial investment of $10–20 million. Once the facility is operational, it will cost $5–7 million a year to add containers as storage needs grow and to maintain the facility. These costs are above and beyond the contributions that electricity consumers already have made into the Nuclear Waste Fund, established to pay for used fuel disposal. As a result of DOE's default on its 31 January 1998 obligation to begin moving used fuel from nuclear power plants, utility customers may have to pay an additional $5–7 billion for used fuel management (assuming the repository is available beginning in 2010, which now looks less likely).

The spent fuel or high-level waste from reprocessing must eventually be sent to a permanent storage facility or 'repository'. The fuel waste is, however, just one of several types of waste that arise from operating a nuclear reactor and each must be dealt with according to its activity and nature.

5.11.3 Waste management

There are three broad categories of nuclear waste arising from reactor operations and other parts of the nuclear cycle.

- Low-level waste contains relatively low levels of radioactivity, which arises from operations associated with radioactively contaminated material, decommissioning and clean-up of nuclear sites, as well as non-nuclear industries. Very low-level waste is a subcategory of low-level waste which includes slightly contaminated clothing and items that come from places such as nuclear medicine wards in hospitals, research laboratories and nuclear plants. Very low-level waste contains only small amounts of radioactivity and can be treated like ordinary rubbish in controlled quantities. These definitions are currently subject to review and consultation.
- Intermediate-level wastes mostly come from the nuclear industry. They include used reactor components and contaminated materials from reactor decommissioning. Typically these wastes are embedded in concrete for disposal and buried.
- High-level waste generally refers in the UK to a waste form in which highly radioactive fission products, formed as a by-product of reprocessing spent fuel, are 'vitrified' – immobilised in glass. It is generally long-lived and gives off significant quantities of heat. Elsewhere the term may refer to spent nuclear fuel from nuclear power plants. The fuel is initially stored in large water-filled pools. The water provides shielding from the radiation and cooling to remove the heat, which continues to be generated by the radioactive material in the spent fuel. After several years, when the radioactivity and its associated heat have diminished, the fuel is transferred to medium-term storage near the nuclear power plants.

There is no consistency in the definitions of these waste categories from country to country or even within a single country. For example, in the USA the US Nuclear Regulatory Commission and US Department of Energy use different definitions of waste. Among the variations:

- The USNRC has classes of low-level waste (A, B, C and greater than Class C) but the USDOE does not have classes of low-level waste.
- The USDOE defines transuranic waste as waste with a concentration greater than 100 nanocuries per gram of alpha emitting TRU, while the USNRC has higher concentration limits for americium-241 (3,500 nanocuries per gram) and curium-242 (20,000 nanocuries per gram).
- The USDOE does not have a low-level waste – long lived classification.

In the UK, radioactive waste is currently categorised according to its heat-generating capacity and its activity content. The categories are:

- High-level or heat-generating wastes (HLW) are wastes in which the temperature may rise significantly as a result of their radioactivity, so that this factor has to be taken into account in designing storage or disposal facilities. IAEA guidance is that HLW thermal power exceeds about 2 kW/m^3.

- Intermediate-level wastes (ILW) are wastes with radioactivity levels exceeding the upper boundaries for low-level wastes, but which do not require heating to be taken into account in the design of storage or disposal facilities. IAEA guidance is that ILW thermal power is below about 2 kW/m^3.
- Low-level wastes (LLW) are wastes containing radioactive materials other than those acceptable for disposal with ordinary refuse but not exceeding 4 GBq/t of alpha or 12 GBq/t beta/gamma activity.
- Very low-level wastes (VLLW) are wastes which can be safely disposed of with ordinary refuse (dust-bin disposal), each 0.1 m^3 of material containing less than 400 kBq of beta/gamma activity (4 Bq/cc) or single items containing less than 40 kBq of beta/gamma activity. The categorisation of VLLW is a working practice adopted by the Environment Agency.

Low-level waste is permanently stored in the UK at a shallow burial site called Drigg, sited adjacent to BNFL's Sellafield site in Cumbria. When the site first opened, low-level waste was buried in it directly. Since the beginning of the 1990s that has changed: the waste is placed in drums which are filled with cement to immobilise the waste. The drums are then placed in concrete-lined trenches. Drigg is unlikely to be large enough to accept all waste that will arise from the existing nuclear programme, especially when large-scale waste arises from the decommissioning and dismantling of the power stations and associated equipment. There have also been concerns that in the very long term Drigg is vulnerable to water incursion from sea-level rise. Its future is under discussion in the light of these two factors.

The fate of intermediate- and high-level waste is still under discussion. It had been assumed that it would be buried in an engineered storage facility deep underground and in the 1980s Nirex, an agency of the nuclear industry, investigated a number of sites to find the most stable geological formation where it could be stored. However, plans to build a deep rock laboratory under Sellafield – the site chosen by Nirex – were cancelled and the UK was left without a storage site. As a result a new body, the Committee on Radioactive Waste Management, was set up to re-examine options for waste management, with input from the public. By the end of 2005 the Committee had narrowed its choices to deep or shallow burial in an engineered store. However, the Committee's remit did not extend to finding a site and that process has still to be started.

Before the waste is sent for disposal it may undergo many types of processing, including the cementation and vitrification referred to above. The aim is to immobilise the waste and reduce its volume as far as practicable. This may include the following.

- Conditioning, which may include the conversion of the waste to a solid waste form, enclosure of the waste in containers and, if necessary, providing an overpack.
- Immobilisation. This may include conversion of waste by solidification, embedding or encapsulation. Immobilisation reduces the potential for migration or dispersion of radionuclides during handling, transport, storage and/or disposal.
- Adding an overpack – a secondary (or additional) outer container for one or more waste packages, used for handling, transport, storage or disposal.

The UK government and the devolved administrations have initiated a review of the long-term solid low-level radioactive waste (LLW) management policy. Many nuclear sites and facilities are now moving into the decommissioning phase, to be managed by the Nuclear Decommission Authority (NDA), and it has been recognised that a very large volume of LLW will arise during the next few decades. A wide range of activities involving the use of radioactive substances generate this LLW. They include the operation of nuclear reactors, the operation of nuclear fuel processing facilities, the decommissioning and clean-up of nuclear sites and non-nuclear industrial activities, including the medical use of radioactive materials and research and educational activities.

The defined activity range for LLW is large, spanning about five orders of magnitude in terms of Becquerels per gramme. Not least because of this, it has been managed in a number of ways in the past: using the low-level waste repository at Drigg in Cumbria, using various forms of disposal on the nuclear site on which the waste was generated, using controlled burial to landfill and, for small quantities of very low-level waste, through disposal with ordinary refuse to landfill.

Potentially the LLW expected to arise in future could fill the remaining Drigg disposal facility capacity a number of times over. This raises the question as to whether this national asset should be used to take large quantities of very low-activity waste or whether there are alternative, more cost-effective ways, in which such waste might be more appropriately but equally safely managed. Furthermore, does it really make sense to dig up very large volumes of very low-activity waste and transport them over potentially large distances for burial at some other location?

5.12 Life cycle analysis

The complexities of the nuclear fuel cycle raise the question of how much carbon dioxide is produced during fuel preparation and spent fuel management. Some studies have been carried out to address this issue for nuclear as well as for other forms of power generation.

The US-based Center for Renewable Energy Technologies summarises the issues that may be included in such a study as follows.

- **Coal.** Energy for mining, methane released from coal beds, energy to transport coal, energy to build plants, combustion emissions, energy to run desulphurisation equipment and dispose of waste.
- **Oil.** Energy to drill, transport and refine petroleum, methane released by drilling, CO_2 from flared gas, energy to build plant, combustion emissions, energy to operate desulphurisation equipment and dispose of waste.
- **Natural gas.** Energy to drill, pipe and refine natural gas, methane escaping from pipelines, energy to build plant and combustion emissions.
- **Geothermal.** Energy to build plant and pipe system, emissions from reservoir, if any.
- **Wind.** Energy to build wind farm.

- **PVs.** Energy to fabricate silicon (or remelt semiconductor scrap) and manufacture PV equipment, energy to manufacture batteries, if used.
- **Fuel cells.** Energy to fabricate fuel cell, energy to produce hydrogen fuel, carbon released by the removal of hydrogen from natural gas, gasoline or other hydrocarbon feedstocks.
- **Biomass.** Producing fertiliser, if any, energy to cultivate biomass, if any, energy to collect and transport biomass, energy to build the facility, combustion emissions, energy to dispose of waste and 'negative' emissions from CO_2 absorbed during biomass growth cycle; may also include 'negative' emissions from avoiding combustion and rotting of fuel.
- **Nuclear.** Energy to mine, concentrate, convert, enrich, transport and (outside the United States) reprocess uranium, energy to build and operate reactor and energy to transport and store radioactive waste.
- **Hydropower.** Energy to clear land, net emissions from permanently lost CO_2-absorbing biomass, energy to build dam and CO_2 and methane from rotting biomass in reservoir.

It is also possible to include the energy required to decommission power plants or recycle equipment. The type of energy used to accomplish these tasks will affect the level of emissions, as will the absolute quantity of steel, cement, aluminium, etc. Note that cement production itself emits CO_2 and that aluminium production releases carbon tetrafluoride, a potent greenhouse gas.

With so many differing inputs to assess, the results of the analysis may vary greatly depending on the assumptions made.

Three versions of such an analysis were compared by the World Nuclear Association and Australia's Uranium Information Centre. These analyses were carried out by Japan's Central Research Institute of the Electric Power Industry, which published life cycle carbon dioxide emission figures for various generation technologies, by Sweden-based utility Vattenfall, which in 1999 published a popular account of life cycle studies based on the previous few years' experience and its certified Environmental Product Declarations (EPDs) for Forsmark and Ringhals nuclear power stations, and by Kivisto, who in 2000 reported a similar exercise for Finland. They show the following CO_2 emissions:

g/kWh CO_2	Japan	Sweden	Finland
Coal	975	980	894
Gas thermal	608	1170	—
Gas combined cycle	519	450	472
Solar photovoltaic	53	50	95
Wind	29	5.5	14
Nuclear	22	6	10–26
Hydro	11	3	—

The Japanese gas figures include shipping liquefied natural gas (LNG) from overseas, and the nuclear figure is for boiling-water reactors. The Swedish nuclear figure assumes 80 per cent of uranium enrichment is by centrifuge; the Finnish range of figures encompass both 100 per cent centrifuge and 100 per cent diffusion.

If extremely low-grade ores are envisaged in the nuclear calculation, the figure would rise by a further 1 per cent in line with the energy inputs.

Sources:

Kivisto A. (1995) *Energy payback period and CO_2 emissions in different power generation methods in Finland*

Vattenfall (1999) *Vattenfall's life cycle studies of electricity* (also energy data 2000)

Chapter 6

Privatising the UK nuclear industry

6.1 The UK electricity system

For 40 years, between the 1950s and the beginning of the 1990s, the UK's electricity supply system was a monopoly that was state-owned mainly through the Central Electricity Generating Board, which came into being shortly after the Second World War. However, all that changed in the late 1980s when the Conservative government, led by Margaret Thatcher, decided that the power industry and other state industries should be in the private sector. After several successful privatisations, including British Gas and British Airways, the government decided to take a similar route with the electricity supply industry.

At that time the electricity industry in England and Wales consisted of the Central Electricity Generating Board (CEGB), which was responsible for generation and high-voltage transmission, and 12 local Area Boards responsible for distribution to domestic and commercial customers. In Scotland there were just two companies, the South of Scotland Electricity Board and the North of Scotland Hydro Board, responsible for generation, transmission and distribution. The Northern Ireland Electricity Board was similarly 'vertically integrated' in Northern Ireland.

Following extensive discussions in the late 1980s it was decided that the CEGB's functions in England and Wales should be split. All the power stations were to be allocated to two generating companies. The high-voltage transmission network, across which electricity from power stations is transported in bulk, was to be turned into a single company, later to be known as the National Grid. The 12 area electricity boards that supplied power to customers at domestic and commercial voltages were turned into separate companies known as distribution network operators (DNOs).

6.2 Regulated and unregulated businesses

In the new industry, neither the high-voltage network operated by the National Grid nor the DNOs generated or sold power to customers themselves. Instead, their function was to maintain and operate the power networks and manage the flow of power from

Panel 6.1 Transmission and distribution companies

Transmission companies

These companies are responsible for maintaining and operating the high-voltage transmission network that carries large amounts of power from the generators to the distribution networks. There are three transmission licence holders in Great Britain – Scottish Power and Scottish & Southern Energy in Scotland, and National Grid and England and Wales. National Grid operates the network across Great Britain but only owns the network in England and Wales.

Distribution network operator companies

See Figure 6.1 for the owners and operators of the network of towers and cables that bring electricity from the high-voltage transmission network to homes and businesses in each region.

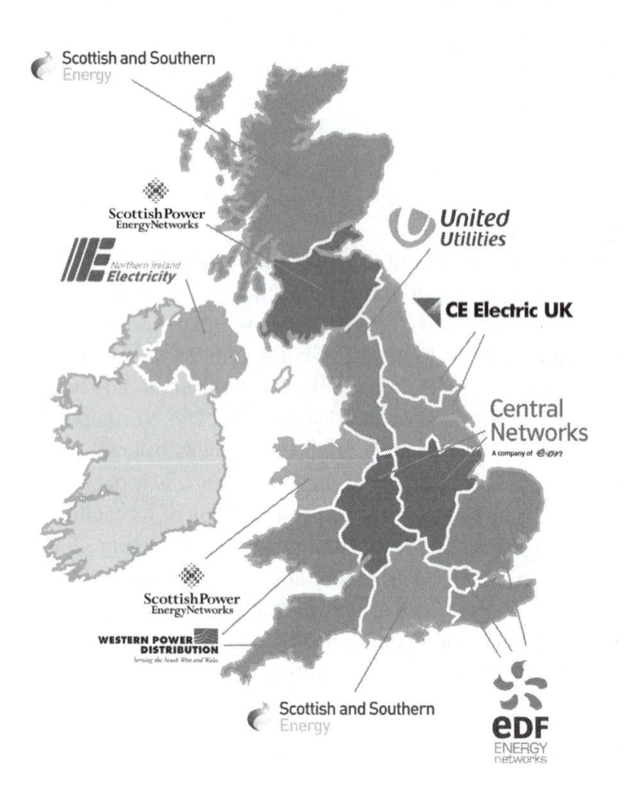

Figure 6.1 Distribution network operator companies [published with permission of the Energy Networks Association (ENA)]

the generating companies, along the wires, to users. Under the CEGB, domestic and commercial customers had previously had no choice but to buy power from their area electricity board. However, after the reorganisation consumers could instead buy their electricity from any one of a new group of 'suppliers'. The suppliers might not own any power generating stations (although some own them now, either directly or indirectly) but instead would buy bulk electricity from the generating companies and sell it on to users. Major users buying electricity in bulk for their own use – such as large industrial users – would be able to bypass the electricity suppliers and choose from the competing generating companies.

This separation of the physical assets of the system from the commodity sold across it enabled competition to be introduced in two parts of the system. Generators would compete with each other to sell power to the supply companies (or to a few very large users who buy similar amounts of power directly). The supply companies would compete to sell power to domestic and commercial consumers. It was left to the market to ensure that these companies charged the lowest price for their electricity. No person or organisation had the responsibility to ensure 'security of supply' (i.e. that there would be enough generating capacity to meet demand in the present and future). Instead, it was also left to the market to provide signals in the form of changing prices and a 'forward price curve' (as buyers predicted electricity would become more scarce). A rising forward price curve would show that supplies of electricity would become scarce and would, for example, prompt generators to build additional generating capacity when it was required.

In comparison, the National Grid and the DNOs would become 'regulated' businesses. The charges they would make for their services, and the financial management of their business, would be set within a framework agreed by a government-appointed regulator (in this case the Office of Gas and Electricity Markets, now known as Ofgem).

In Scotland the two companies' activities were split in a similar way and the new companies were able to compete with the England and Wales companies, but the two markets remained separate. This changed in 2005 when a single market system was introduced (known as the British Electricity Transmission and Trading Arrangements, or Betta).

The split into generation, transmission, distribution and supply is now widely accepted and has been used in numerous other electricity supply industry deregulations and privatisations worldwide.

Since the UK's industry was sold off there has been plenty of consolidation within the sector and some generators and suppliers have merged or been bought by other companies. In fact, some assets have changed hands several times. Now there are UK, European and US utility companies active in the market. The players in the market at the end of 2005 are shown in the Panels 6.2 and 6.3.

6.3 Nuclear issues

The electricity privatisation process was not without other hurdles, but by far the biggest and longest running was the issue of how to handle the UK's nuclear stations.

Panel 6.2 Power generating companies

These are the companies responsible for generating the energy used in homes and businesses.

Major power generators

British Energy plc	www.british-energy.com
British Nuclear Fuels plc	www.bnfl.com
Drax Power Ltd	
EDF Energy plc	www.edfenergy.com
E.ON UK plc	www.eon-uk.com

International power generators

RWE npower plc	www.rweinnogy.com
Scottish and Southern Energy plc	www.scottish-southern.co.uk
ScottishPower plc	www.scottishpower.plc.uk

Smaller generators and co-generators

AES	
Corby Power Ltd	
Corus Group plc	www.corusgroup.com
Deeside Power Development Co Ltd	www.deesidepower.com
Ecotricity	
Fibrowatt Ltd	www.fibrowatt.com
Humber Power Ltd	www.humberpower.co.uk
Medway Power	
Natural Power Company Ltd	www.naturalpower.com
Optimum Energy Ltd	
Saltend Cogeneration Co Ltd	
Seabank Power Ltd	
Slough Heat and Power Ltd	www.sloughheatandpower.co.uk
South Coast Power Ltd	
Summerleaze Ltd	www.summerleaze.co.uk
Tanaris Ltd	www.tanaris.com
Teesside Power Ltd	www.teessidepower.co.uk
Tokyo Electric Power Co Inc	www.tepco.co.jp/en/in dex-e.html
Western Power Generation Ltd	www.westernpower.co.uk

The confidence of planners in the late 1980s that the nuclear stations could simply be sold with the other generators was quickly found to be misplaced. The nuclear fleet was eventually – a decade later – transferred to the private sector, but its early success as a private company was short-lived.

Panel 6.3 Electricity supply companies

These are the companies who supply and sell electricity to the consumer.

Major electricity suppliers

British Gas
Centrica
EDF Energy
London Energy (EDF Energy)
Manweb (Scottish Power)
npower
Powergen
Scottish Hydro Electric (Scottish and Southern Energy)
Scottish Power
Scottish and Southern Energy
Seeboard Energy (EDF Energy)
Southern Electric (Scottish and Southern Energy)
Swalec (Scottish and Southern Energy)
SWEB Energy (EDF Energy)

New and specialist electricity suppliers

Atlantic Electric and Gas
Basic Power
Countrywide Energy
Ecotricity
Good Energy
Green Energy UK

6.3.1 Waste liability

One of the most important problems with privatising the UK's nuclear stations was the cost of managing radioactive nuclear waste.

Over its lifetime a nuclear plant has several sources of radioactive waste. Some of them arise during the operation of the plant. That might range from simple rubbish that has been used inside the 'active' areas of the plant to highly active used (spent) fuel that has been unloaded and replaced with fresh fuel. The second major phase of waste production would eventually arise when the plants were 'decommissioned'. Decommissioning is a complex process in which the fuel is removed from the reactor and the reactor structures and buildings themselves are dismantled, removed, packaged and taken to a storage site (see Chapter 5).

The costs of decommissioning a nuclear station were extremely uncertain in the late 1980s, because at that time the only experience of the process was from very small research reactors. Knowledge of the process and experience are now more extensive, but it is still far from a familiar or routine process and the costs in the UK consequently have many uncertainties and are estimates at best. Technically, many of the necessary techniques were being developed and smaller decommissioning projects had provided useful experience. But that did not translate into the detailed financial data that would be required to give confidence to potential investors in a privatised electricity supply industry.

In other countries the same uncertainties existed but, in some, the financial arrangements for dealing with them were much better defined.

In the US, for example, the ultimate responsibility for dealing with spent nuclear fuel and waste from the reactors lies with the government. The reactor operators pay the government a levy on each unit of power they sell; the government in return has to find and prepare a repository for the waste.

In fact, the theory in the US has been simpler than the practice. The government was due to 'take title' to the operators' spent fuel at the end of the last century but the repository had not been finalised (although a site at Yucca mountain was undergoing the long process of technical and regulatory qualification). Operators had made interim arrangements to store fuel and taken legal action against the government to recover their costs. Nevertheless, the legal position is clear.

6.3.2 The UK approach to waste

In the UK, however, the position was different. In the UK it was assumed that the nuclear power industry would continue to grow and that uranium fuel would become scarce and expensive. The government therefore took the decision that the uranium and plutonium in spent fuel were an asset and should be removed from the fuel by reprocessing. That would involve mechanical and chemical reduction of the fuel into its component parts and chemical extraction of the uranium and plutonium for reuse. What is more, the government reasoned that this was an area where British expertise was also a national asset, and one that could be offered to other nuclear operators overseas.

Britain's military nuclear history was consistent with this choice but other countries also took this view: Germany, Switzerland and Japan were among the countries who signed contracts with British Nuclear Fuels to reprocess spent fuel, sometimes as an interim measure as they were considering whether they would need their own reprocessing plants. France took a route similar to the UK in building reprocessing facilities and offering its services to other operators.

The decision to follow the reprocessing route for its power reactors locked the UK and the CEGB into major investments and international contracts that robbed it of any opportunity to change direction. In the original UK context, these arrangements were made between government-owned industries. But the possibility of privatising the nuclear power stations with the rest of the electricity supply industry meant the potential costs had to be fixed and allocated to the various companies that would succeed the CEGB. That process has taken nearly 15 years, while the UK's nuclear investment has been in and out of privatisation.

6.3.3 Unquantifiable costs

At the time the privatisation began, the financial liabilities were unquantifiable. But since there were no state bodies set up to deal with them the nuclear plant would carry that liability into the private sector.

The nuclear industry was not unique in having 'legacy' liabilities in this way. British Coal was broken up and its liabilities transferred to the Department for Trade and Industry in 1998, under the terms of the Coal Industry Act of 1994. But the coal liabilities were relatively familiar and quantifiable – and they were firmly in the public sector, and staying there (see Panel 6.4).

Panel 6.4 Liabilities

The nuclear industry's liabilities are not the first ones taken on by the government, nor will they be the last. Carbon capture and storage is a fledgling industry, for example, which will take processes already in use in the oil and gas industry in which carbon dioxide replaces the fossil fuel in undersea deposits. The process is being investigated as a method of storing carbon dioxide produced from power stations or large industrial sites. Companies developing the process have already made it clear that government will be expected to take on the long-term liabilities of the stored carbon dioxide.

British Coal was broken up and its liabilities transferred to the Department for Trade and Industry in 1998, under the terms of the Coal Industry Act of 1994. That meant the DTI took on liability for compensation claims for around three quarters of a million miners who suffered from respiratory disease (chronic bronchitis and emphysema) resulting from the dusty conditions or from vibration disease (Hand Arm Vibration Syndrome) caused by using vibrating tools. The DTI eventually expects to make payments totalling £7.5 billion to such claimants.

In addition the Coal Authority, set up in 1994, licenses mining operations but is also responsible for dealing with historic liabilities relating to its ownership of abandoned coal mines. This includes making safe ground collapses and the monitoring of ground water levels and mine gas emissions, and for this it received a typical grant in aid of £20 million in 2000.

But after a decade of investigation and shifting responsibilities, nuclear's liabilities are revealed to be of a different order. The Nuclear Decommissioning Agency estimated the clean-up cost at £56 billion when it issued a draft strategy in August 2005 and was clear that there were likely to be further increases.

At that time some of the costs for major parts of the programme – including dealing with plutonium and cleaning up Sellafield had not been well defined and there were questions over planned income from commercial activities such as the THORP reprocessing plant, income from which had been expected to defray decommissioning costs.

The NDA was set up following the Energy Act 2004 to take over the UK's civil nuclear liabilities. It has responsibility for 20 sites including Sellafield, Dounreay, the Magnox reactors and various fuel manufacturing and research sites.

Among the liabilities still to be captured are:

- The reprocessing site at Sellafield, which includes many of the higher hazard facilities which are the main priority for clean-up and may represent 60 per cent of contract costs. At both Sellafield and Dounreay (where clean-up plans are more advanced) the precise contents of the facilities were not known in 2005 but the cost of dealing with the higher hazard facilities was expected to be 'very high'.
- Waste: There are no long-term management arrangements in place for high- or intermediate-level wastes. However, the Committee on Radioactive Waste Management has published a final shortlist of options, including interim surface storage and deep geological disposal, possibly in a phased, retrievable fashion. The NDA favours deep storage but says that assessing and preparing a site would take 25–40 years, which could disrupt its plans to bring forward reactor dismantling. On low-level waste, the UK has stores at Dounreay and at Drigg near Sellafield – although Drigg is running out of capacity. Moreover, Drigg's future has been questioned by the Environment Agency which has warned that safety cases for the site 'fail to make an adequate or robust argument for continued disposal'. The Agency is concerned about the possibility of coastal erosion, which may affect Drigg in as little as 500 years, and has called for the removal of long-lived wastes which were dumped before 1985. The NDA proposes a new site, again near Sellafield.
- Plutonium: In the past, plutonium from spent fuel reprocessing has been regarded as an asset – albeit one with zero value. There is growing pressure to reclassify some or all of this as a waste, which could increase NDA's liabilities by a further £5–10 billion. Similarly, reprocessed uranium and several uranium-containing by-products of fuel manufacture (notably depleted uranium) could all be reclassified as waste.
- THORP and MOX: The NDA inherited these plants from BNFL as assets, and revenue was supposed to help pay for wider clean-up activities. However, both are in deep trouble. THORP remains shut down following a serious leak in 2005. The NDA will present its conclusions on THORP 'when current technical issues are clearer' – but says the decision on restarting the plant is a matter for the government. Reprocessing contracts run out in 2010 and it will be necessary to develop alternatives, such as storage of AGR fuel, in any case. Moreover, THORP's operations are linked to other activities on the site and may need to be halted if, for example, the site's troubled vitrification plant struggles to meet regulatory targets to

reduce the stockpile of highly active liquid wastes. The MOX plant has also suffered technical problems. The NDA said it intended to 'take a close interest' in the plant's commissioning and would discuss its future with the government if it proved unable to meet production targets.

- Contaminated land: There is thought to be radiological contamination of 20 million cubic metres of land under the Sellafield site. Dounreay and Springfield, and probably the Magnox sites, have smaller volumes of contamination. The NDA has not decided whether this should be removed, treated *in situ*, or left in place. It also says there is 'no easy solution' to the persistent problem of small radioactive particles on the coastline near Dounreay.
- Clean-up standards: Before 2004 it was assumed that most nuclear sites would eventually be returned to the 'green field' state and released for unrestricted use. Following a review in 2004, however, the NDA says there could be 'a variety of site end states' – including indefinite institutional control, as envisaged at Sellafield. This may offer the 'opportunity to reduce costs', the NDA says.

The nuclear liabilities had to be laid bare for potential investors, or there would be no nuclear privatisation. In 1988 and 1989 the Department of Energy and its advisors in the privatisation process began to try to come to grips with the waste liabilities. The extent of the problem was pure speculation at this stage, but it would clearly mean large numbers, not just for the waste itself but for issues around it. For example:

- There was no permanent site for storing intermediate-level waste (ILW). A company funded by the nuclear industry, Nirex, had spent a decade or more searching for a potential site for a deep repository, but that programme had fallen through and a new search had to begin.
- Waste costs were rising due to other environmental legislation that had halted, for example, waste dumping at sea.
- The CEGB was liable for its share of the decommissioning costs of BNFL's facilities, which had been and were being used to reprocess nuclear fuel. This took the form of so-called cost plus arrangements with BNFL for supply of fresh fuel and management (i.e. reprocessing and disposal) of spent fuel, in which the electricity company paid all the costs of the process – whatever they may have been – plus an additional payment. Among the charges that would be included were the cost of decommissioning the plant at BNFL's Sellafield site, used to reprocess spent fuel, plus all the associated facilities – another unquantified cost.
- It was very unlikely that the CEGB or a successor could cancel the reprocessing contracts. Because the assumption was that fuel would be reprocessed, fuel for some UK reactor designs was not designed for long-term storage, so without a new programme of research and some special measures there was no other option. Fuel from the UK's PWR could have been disposed of directly, but the CEGB had

Figure 6.2 Storage pond in Thermal Oxide Reprocessing Plant (THORP). Sellafield, United Kingdom [published with permission of the Nuclear Decommissioning Agency (NDA)]

forward contracts for PWR fuel management with BNFL that formed an important part of the BNFL's case for building a new reprocessing plant, the Thermal Oxide Reprocessing Plant (THORP). Another part of the case for THORP was that it had contracts to reprocess fuel for overseas nuclear operators.

- The CEGB's plan for dismantling its nuclear stations involved removing non-nuclear or lightly contaminated secondary structures and leaving a compact controlled building containing the nuclear core and major components. That would be maintained for 130 years – referred to as 'safestore'. The plan had some technical and financial merits: radioactivity in the components would decay significantly over a century and the eventual dismantling of the station would be safer and simpler. Financially, a relatively low investment now could grow to fund the decommissioning process in the future. But for investors it was yet another open-ended and apparently unquantifiable liability.
- Was there, in practice, any decommissioning fund that could be used to finance the work when it was required? Ideally, a small levy on electricity prices over the lifetime of the plant would be held in a separate fund. But in public ownership the CEGB's finances were a function of the Treasury. Decommissioning liabilities would have arisen as current account costs far in the future, not as a single payment to provide a fund for future use by the putative owners of the nuclear stations.

6.3.4 In and out of privatisation

Before the start of the privatisation process, the fact of nuclear's waste liability was known. Indeed, when the UK's power stations were allocated to potential generating companies that would follow the CEGB the split was weighted to take account of this.

There were expected to be two generating companies, informally known as Big G and Little G. Big G – later to become National Power – included all the nuclear stations and some conventional generation, in a company that was thought to be big enough to handle the nuclear liabilities. Little G – later Powergen – had conventional generation only. But it did not work because it was never clear that National Power would be big enough to take on the nuclear stations. In fact, it quickly became clear that investors would not take any stake in a company with such uncertain and potentially enormous liabilities.

The crunch came in the second half of 1989. At first it had been thought that the biggest liabilities were in the Magnox stations which were near the end of their lifetime, whereas the AGRs and Sizewell B were economically more attractive. In July 1989, Energy Secretary Cecil Parkinson announced that the Magnox stations would stay in public ownership. But even removing the Magnox stations did not resolve the hole in the nuclear account. Cecil Parkinson was replaced as Secretary of State for Energy by John Wakeham, and just four months after the Magnox withdrawal the rest of the nuclear stations were pulled out of the privatisation.

The nuclear stations were regrouped into a nuclear-only company, to be called Nuclear Electric. It was announced that Nuclear Electric would stay in public ownership, while Powergen and a slimmed-down National Power would be sold off.

The privatisation in Scotland had been following a parallel course to that of England and Wales, but in Scotland the two AGR stations, at Torness and Hunterston B, which together represented 25 per cent of the generating capacity held by the original Scottish electricity boards, had already been placed in a single company, while Hunterston A Magnox station was approaching the end of its useful life. The country's nuclear power stations were also withdrawn from the privatisation after a separate statement to the House of Commons by Malcolm Rifkind, Secretary of State for Scotland. The Secretary of State said 'the full flotation could not be successfully achieved without wide-ranging and unequivocal indemnity from Government for future cost escalations. We do not consider that it would be in the public interest to confer such indemnity, and have therefore decided that the nuclear enterprise in Scotland should remain in the public sector.'

Since it had already been proposed that the Scottish nuclear plants should be owned by a separate company, Scottish Nuclear Ltd, it was decided that the company would now remain publicly owned and would sell its output to the new Scottish supply companies, Scottish Power and Hydro Electric.

6.3.5 The end of expansion

The withdrawal also called time on Britain's plans for expanded nuclear generation. There had been years of debate in the late 1970s and early 1980s over whether another

Panel 6.5 Parliamentary debate – nuclear privatisation

The decision to withdraw the nuclear stations from privatisation was a watershed in the industry's history. The announcement was made in the House of Commons on 9 November 1989 by John (now Lord) Wakeham, who had been appointed Secretary of State for Energy by Margaret Thatcher earlier that year. The following gives a flavour of the debate that followed his statement:

> The Secretary of State for Energy (Mr. John Wakeham): With permission, Mr. Speaker, I should like to make a statement about electricity privatisation and nuclear power.
>
> The Government remain determined to complete electricity privatisation, with all the benefits it will bring, in the lifetime of this Parliament. The preparations for the new system have brought issues into the open. In particular, attention has been focused on the costs of nuclear power. The Government have for some time recognised that our nuclear power is more costly than power from fossil-fuelled generating stations....
>
> The Government told the House on a number of occasions during the passage of the Bill that the arrangements for nuclear power would strike the appropriate balance between the interests of the taxpayer, the electricity consumer and the shareholder. In the event, unprecedented guarantees were being sought. I am not willing to underwrite the private sector in this way. The Government attach the highest importance to the successful completion and operation of Sizewell B in order to maintain the PWR option in the United Kingdom.
>
> The distribution companies need to be clear what their obligations will be for a reasonable period ahead. The Government will wish to review the prospects for nuclear power as the Sizewell B project nears completion in 1994.
>
> The price of the nuclear company's electricity will be set at a level consistent with its earning a return appropriate to public sector bodies. Since the price of nuclear electricity is likely still to be somewhat higher than that of fossil-fuelled electricity, the fossil fuel levy will be used to share the additional cost over all electricity suppliers.
>
> Frank Dobson (Holborn and St. Pancras): ... the Government insisted that nuclear power stations should be privatised. In the words of the then Secretary of State for Energy in April this year, private ownership of nuclear stations would get them away from political influence and make them better managed, better off and more accountable....
>
> So much for competition. We know that whoever decides the price of this nuclear power in future, it will certainly not be the customers who, we were told, would be of paramount importance.
>
> ... this statement spells the end of the Government's commitment to an expanded programme of nuclear power
>
> Tony Benn (Chesterfield): Is the Secretary of State aware that, in his statement, he has simultaneously destroyed the case for privatisation, undermined the case for nuclear power by admitting that it is much more expensive, and retrospectively destroyed the Government's war against the mining industry, which he today admitted can generate electricity much more cheaply than the nuclear industry can?
>
> Michael Jack (Fylde): may I ask him on behalf of the 3,000 nuclear workers in my constituency when the plans for the future of the Magnox stations will be available, when the details of the remedial works on the AGRs will be available,

and when he will be able to advise British Nuclear Fuels plc about the future of its £ 123 million investment in the PWR fuel plant?

Malcolm Bruce (Gordon): … what obligations have been made to Westinghouse for the PWRs that are now being cancelled, and the cost that will fall on the taxpayer as a result of that decision?

Sir Ian Lloyd (Havant): … May I welcome the Government's decision to maintain the nuclear option, as all evidence reaching us about the greenhouse effect suggests that there is no other realistic option?

Will the price be essentially determined by fossil fuel generators, even when their prices rise, as they are bound to do, when vast additional costs are imposed on fossil fuel stations to meet clean air conditions? Finally, does the revision of decommissioning costs – which were elaborately discussed in Lord Marshall's paper on 6 December, and which amount to between £8 billion and £12 billion – lie at the heart of my right hon. Friend's decision to revise the plan?

Mr. Wakeham: One of the difficulties with which we have had to deal has been that the costs of nuclear power remained hidden throughout nationalisation, and it was only the preparations for privatisation that brought them to light.

Sir Trevor Skeet (Bedfordshire, North): … when the price of coal rights itself and the price of oil goes up in about five or six years, will not nuclear energy come into its own and blossom forth in the economy?

Mr. Dennis Skinner (Bolsover): … now that we have reached a watershed in nuclear power and other forms of energy, it would make a lot of sense to stop the pit closure programme, to say that the CEGB's coal allocation should be increased from 60 million tonnes to 75 million tonnes, that coal imports should be stopped and that the Associated British Ports (No. 2) Bill should be dropped?

Mr. Peter Rost (Erewash): … nuclear electricity is uneconomic and produced at a high cost only in Britain, not in those many other countries that have done the job sensibly …

Donald Dewar (Glasgow, Garscadden): … 60 per cent of all electricity sold in Scotland comes from nuclear capacity …

Does the Secretary of State now agree with the Department of Energy that, if that is done, the true cost of nuclear power is at least twice that of conventionally produced power?

nuclear programme was required, and if so whether the British-designed AGRs should be favoured, or whether the UK should take on the US-designed PWRs or BWRs that were becoming familiar in other countries. In the end, a series of four PWRs had been ordered from Westinghouse in the USA, in the teeth of vocal public opposition, and the plan had been tested in a planning inquiry for the first plant, at Sizewell, that had gone on for over two years.

By 1989 Sizewell B was finally under construction, while a second PWR, Hinkley Point C, was already undergoing a public inquiry, which in fact was under way at the time of the withdrawal. Two further plants were planned at Wylfa (designated Wylfa B) and Sizewell (Sizewell C). But when the nuclear stations were removed from the privatisation the remaining three reactors in the series were cancelled. That decision later had implications for Sizewell B as it had to bear all the so-called first-of-a-kind licensing and design costs that were intended to be shared across the four-plant series.

6.3.6 *Fallout from the decision*

The following year an inquiry by a government committee, the House of Commons Select Committee on Trade and Industry, examined the debacle. It said that 'After decades of support for the expansion of the nuclear industry, the Government Statement on 9 November 1989 represents a complete reversal of policy. Additionally, nuclear power, previously promoted as a cheap source of energy, has, since the possibility of privatising the industry, become expensive. The Committee has undertaken this inquiry to discover the true cost of nuclear power and why the Government had not recognized this more quickly.' It discussed the costs associated with different sources of nuclear power and concluded that, 'while there were some factors which have impacted upon the costs, such as the tightening of regulatory standards and the rising cost of capital, the largest factor was a bias by the Central Electricity Generating Board (CEGB) in favour of nuclear power.' The Committee said it held 'the Department of Energy partly responsible for this in that they made no attempt to obtain realistic costings from the CEGB until motivated by privatisation'.

It recommended that 'comprehensive information should be published on the fossil fuel levy, nuclear generation costs and the accounts of Nuclear Electric and SNL; accurate estimates should be made regarding the cost of completing and operating Sizewell B and extending the life of Magnox reactors; and rates of return should take into account the greater risks involved in nuclear energy'.

6.3.7 *Nuclear Electric*

When the non-nuclear generators were sold off in 1991 Nuclear Electric remained in government hands. But it was clear from experiences overseas that it was not impossible for nuclear stations to be owned and operated by private sector companies. In the USA, at least, some nuclear stations changed ownership in the 1990s, so there were willing buyers. Nuclear Electric took it as its aim to show that the UK nuclear stations could be attractive to the private sector.

There were two immediate areas where major gains might be made. Clearly one was the waste liability. But the other was the same imperative that impels any company: increase income and reduce operating costs.

6.3.7.1 Increasing income

The operating history of the UK's nuclear plants had been mixed, to say the least. While the Magnox stations had been fairly reliable performers they produced relatively little power, and although their closure date had been delayed they would reach the end of their lifetime within a decade or two. The AGRs were much younger – but they had performed extremely poorly (see Chapter 5 for details).

The biggest problem at the AGRs was the refuelling process. They had been designed for continuous operation, so that they could be refuelled while the reactor was in operation, making them highly efficient. But it had not worked – and worse, since the reactors were designed to refuel at power, shutting them down to refuel was a long process, as in some cases the reactors not only had to be shut down but the high-pressure primary circuit had to be depressurised before fuel load and repressurised before restart. It meant in some cases weeks out of action.

An immense engineering effort partly solved the problem. Some reactors were brought up to design and could refuel at power. Some did not reach that standard but they could, at least, refuel without depressurising and repressurising the primary circuit. The variation arose because the AGR series has a variety of different versions, designed and supplied by different companies, so solutions to a problem at one reactor can rarely be transferred in their entirety to another reactor. This is a legacy of the UK's approach in the 1960s and it contrasts dramatically with the French nuclear industry, where more than 50 reactors have just a handful of designs, and upgrades or changes at one plant are replicated at all similar plants to ensure they stay in step.

The variation in the UK plants could affect day-to-day maintenance in unexpected ways. Components and spare parts were identified using different numbering systems although many were interchangeable, and operators say it was not impossible to spend thousands on a major replacement component for one plant, only to find that one was already in stock elsewhere.

Other operating costs reductions could include:

- Reducing the bill for operating waste. Materials management programmes that made changes as simple as removing packaging before taking replacement parts into active areas reduce the day-to-day waste arisings.
- Reducing staffing levels. The company reduced its manpower levels by several thousand, at one point running into problems with the nuclear regulator who argued that the company was not retaining enough expert staff.
- Planning maintenance better. Even when reactors are able to undergo on-load refuelling, regular shutdowns of the reactor (known as outages) are required for maintenance. Careful planning and work tracking can reduce the shutdown time, and can also be ready to take advantage of unplanned outages. These can occur due to shutdowns within the reactor or in response to conditions on the grid outside the plant: if the frequency or voltage falls outside a predetermined range, for example, nuclear power plants, like other power plants, will disconnect and shut down as a self-protection. Restart can take several days, and can be an opportunity to complete maintenance tasks.

Nuclear Electric had considerable success in improving operational performance. According to industry magazine *Nuclear Engineering International* it increased load factors at its plants – a measure of how much of their potential generation each plant managed to achieve. The load factors increased from as low as 60 per cent, until in one year its average was over 80 per cent.

According to Nuclear Electric, between 1992 and 1999 the nuclear generating company increased its electricity output by 64 per cent, reduced unit costs by 35 per cent, raised productivity by over 100 per cent, increased operating profit from a loss of £286 million to a profit of £307 million and improved the load factor of its AGRs from 59 to 79 per cent. And with outage and refuelling times dramatically reduced, outages could be better scheduled so that they coincided with periods of low demand, and hence low electricity prices, so revenue increased from electricity sales that had been made.

Nuclear Electric argued that improved performance went 'hand in hand' with improved safety, and in fact the group accident frequency rate for industrial accidents was halved from its 1991 level.

Dealing with the spent fuel and resulting waste liability was more problematic. Negotiations between British Nuclear Fuels and Nuclear Electric continued for many years, during which time the nuclear operator attempted to develop alternative waste management options for its AGRs and Magnox stations, and simultaneously tried to change the contractual arrangements for its PWR fuel so that it would go straight to disposal or to interim storage, as had happened at plants overseas, instead of being reprocessed. But THORP was being completed and started up and BNFL was mindful of the possibility that it was also a potential candidate for privatisation. The two companies could not come to an agreement on a different contract and the existing arrangements remained – a situation that was to be decisive later.

The successes of Nuclear Electric in improving performance were real and significant, but they were also underwritten by a generous subsidy known as the fossil fuel levy, and by an obligation on distribution companies to buy electricity that was not generated by fossil fuel plants.

As its name suggests, the theoretical function of the levy and the obligation were to support any forms of generation that did not rely on coal, oil or gas fuels. In fact it did provide funding for several small renewables projects. In 1989, the Secretary of State for Energy had said, 'there will continue to be the opportunity, as has been made clear previously, for other non-fossil generation to contribute to the non-fossil fuel obligation. In particular, we will maintain the arrangements for the 600 MW special tranche for renewables.' But that was very small beer in comparison with the nuclear industry's nearly 15,000 MW of generation, and almost all the levy went to support nuclear – in theory, to build up a fund that would eventually be used to pay reactor decommissioning costs. The government set the rate of the levy, and it represented 10 per cent of all consumers' bills and about half the income of the nuclear company.

6.3.8 Nuclear Electric to British Energy

In 1994 the Conservative government carried out a 'Review of the Future Prospects for Nuclear Power in the UK', which examined the economic and commercial viability of new nuclear power stations in the UK. The review confirmed the then government's support for the continued operation of existing nuclear power stations. It acknowledged that nuclear power contributed to diversity of supply and to protection of the environment – mainly because nuclear power stations emit practically no sulphur dioxide or nitrogen oxides (the principal ingredients of acid rain) and virtually no carbon dioxide (the principal gas responsible for global warming). It also took the view that providing public sector support for a new nuclear power station would constitute a significant intervention in the electricity market and that circumstances did not warrant such an intervention.

The 1994 review also concluded that moving as much of the nuclear generating industry, with its associated liabilities, as is practicable into the private sector would bring benefits for the industry, electricity consumers and the taxpayer.

Following the review, the nuclear generating industry – which now encompassed Nuclear Electric, Scottish Nuclear and BNFL – was therefore reorganised. The split between the older Magnox and newer AGR and PWR stations, raised during the abortive privatisation in 1989, was revived.

The more modern stations, the seven AGRs and Sizewell B, were now to be transferred to the private sector with their associated liabilities. They were transferred to a holding company called British Energy, whose subsidiary – British Energy Generation Ltd – operated the PWR and five AGR stations previously run by Nuclear Electric Ltd, and the two AGR stations in Scotland previously run by Scottish Nuclear Ltd.

The older series of stations – the Magnox plants, which had only a few years of life remaining – would remain in the public sector with their liabilities. Those Magnox plants still operating – on four sites – were brought together with five Magnox stations that had already been shut down. Initially these stations were the responsibility of a stand-alone company, Magnox Electric plc. However, in March 1998 this was integrated with British Nuclear Fuels plc (BNFL), the UK's government-owned supplier of reprocessing and other nuclear fuel cycle services, which already operated the UK's two oldest Magnox stations (Calder Hall and Chapelcross).

British Energy plc was privatised in July 1996. At the same time, the fossil fuel levy was discontinued.

6.3.9 British Energy in the private sector

At first, between 1997 and 2000, British Energy's performance as a private nuclear generating company was fairly successful.

The company had been sold off on 14 July 1996 and was initially floated at a partly paid offering price of 100 pence per share.

Speaking a year after the sell-off in 1997, the then chief executive Robin Jeffreys said commercialisation had been a success because the issues that had previously prevented the nuclear power stations from being privatised had been addressed. He said privatisation of British Energy had three benefits: the price of electricity to customers was reduced, the government received the proceeds from the sale, and the company was freed from the shackles of state ownership. And he noted that 98 per cent of British Energy employees had taken up shares in the company.

Robin Jeffreys summarised its first year's progress in September 1997, when he pointed out that the share price had doubled over the year to above 200 pence per share. He said, 'That most sceptical group of people – the UK investors – put their hands into their pockets and their money on the table, thus selling BE off into the private sector' and they had been rewarded, as 'British Energy shares significantly outperformed the FTSE All Share Index in a year of strong performance'.

The company was by then one of the UK's largest electricity generators with around 21 per cent market share. Its turnover exceeded £1.8 billion and its market capitalisation was over £1.7 billion with significant cash resources.

As a private or publicly owned entity, Britain's nuclear stations have to operate within the electricity supply market and, like other forms of generation, the characteristics of nuclear as an electricity provider partly determine how operators manage their product over short and long timescales.

In a deregulated market situation, as now obtains in the UK, nuclear's imperative to operate means that it becomes a 'price taker'. This becomes clear when examining British Energy's position in the initial arrangements for the UK market.

In the initial setup of the UK market, some large buyers (including domestic suppliers) came to long-term agreements with power generating companies.

The remainder of the power required in the UK was sold through a mechanism known as the 'pool'. Generating companies would bid to supply power to the pool in half-hourly time slots. The pool had to supply all buyers not covered by long-term contracts. That meant that in peak times, when more power was required, more expensive generating plants would make bids into the pool. The system operator would dispatch power as required from the pool, and would pay all the generators that supplied at the level of the highest bidder that had been called on. Operators of plants with high operating costs and plenty of flexibility could take the gamble of making a very high bid, because they had the option not to supply. For operators like Nuclear Electric, however, the imperative was to operate (wind power suppliers were in the same position). They would therefore bid zero into the pool and would be paid at a low rate for night-time generation and high rate at peak times, hoping that the average would provide an economic return.

The pool mechanism is still used in many countries, but in England and Wales it was replaced by the so-called new electricity trading arrangements (NETA) in 2001, which since 2005 have also covered Scotland (under the name British Electricity Trading and Transmission Arrangements, or Betta). NETA took away the relative certainty of the pool. Instead, buyers were required to obtain all of the power they needed through bilateral contracts with power generators. The National Grid performed final adjustments, known as 'balancing', calling on companies and suppliers who can increase or reduce their demand or supply to fine tune the supply.

NETA went live in March 2001 and it had a dramatic effect on the market. Electricity prices had been expected to drop – after all, it was part of the rationale behind the change in market structure – but the reduction, in some cases, was of the order of 40 per cent.

The change was problematic for all the generators – two generating companies went out of business at this time, and all of them found the trading conditions so hard that the head of one generating company was quoted as saying the 'market was bust'.

British Energy's market position was already weak. Most large generating companies were by now part of large integrated Europe-wide utilities that also had supply companies providing electricity to domestic and commercial consumers. The success of the supply business could balance the hard trading conditions in the generation company.

British Energy had previously attempted to build a secure supply business, by buying Swalec, which supplied power to customers in Wales. The attempt was disastrous. British Energy bought the group in 2000 but sold the business on to Scottish and Southern Electricity a couple of years later. But when British Energy bought

Swalec it came complete with a long-term contract to buy power at fixed prices from a fossil-fuelled power station in Teesside that dated back to the 1990s. When British Energy sold Swalec, Scottish and Southern Electricity refused to take on the Teesside contract, which required British Energy to continue buying a fifth of the power from Teesside at prices that were now around 80 per cent above market value. That left British Energy with a multi-million pound contract to buy power it did not want.

British Energy's other adventures in diversification were not much more successful. Focusing on generation, it had attempted to diversify into non-nuclear power, buying a coal-fired generating station called Eggborough in the Midlands. The station should have allowed British Energy to offer power in the lucrative 'balancing' market. But it was criticised for paying over the odds for that station. Meanwhile, it tried to leverage its nuclear expertise by buying or managing nuclear stations in the USA and Canada.

In September 2002 matters came to a head following the shift to NETA. Electricity prices were far too low to allow the company to cover its costs. At the same time, its generating capacity was far below what had been planned, because several of its stations were out of operation for several months following technical problems. In the six months to September 2002 British Energy made a loss of over £300 million. That included a loss of over £100 million in its decommissioning fund, caused by a falling investment market that hit other big funds like pensions funds equally hard. The ongoing negotiations with BNFL over spent fuel management had not made any progress and British Energy was still committed to paying to reprocess its fuel at BNFL's Sellafield plant.

British Energy's share price, which had reached over £2 at one point after the privatisation, fell by 20 per cent and a further 65 per cent, until it had fallen to almost zero. In September 2002 it withdrew from the stock market.

Nevertheless, British Energy was still the country's largest electricity supplier, at around 20 per cent, and the cost of dealing with its liabilities could only be defrayed if the company existed and had income. The government gave the company a month's grace to work out a rescue plan and eventually decided to bail it out.

6.3.10 What went wrong?

British Energy's spent fuel management arrangements had been under discussion for nearly ten years, since the UK's publicly-owned Central Electricity Generating Board was split into competing companies. British Energy argued now that reprocessing is expensive and unnecessary, that long-term storage is secure and offers fewer non-proliferation concerns, and most of all that it is much cheaper – perhaps half the £400 million per year that it spent on reprocessing. But British Energy entered the private sector saddled with long-term agreements under which all the spent fuel from its nuclear plants would be reprocessed at Sellafield and years of pressure had not convinced BNFL to cancel the contracts.

Part of the problem may have been BNFL's own shaky financial situation and wavering by successive UK governments over whether it should be a public or private company. In the mid-1990s the aim was to follow successful sell-offs of British Energy and parts of the UK Atomic Energy Authority with disposal of most of BNFL as a nuclear services company. That gave BNFL an incentive to retain its British Energy contracts intact.

British Energy continued to press for change but said in its 9 September announcement that new proposals delivered by BNFL on 8 September 'fell short'. Some UK commentators speculated that a better deal from BNFL had been vetoed by the Treasury. British Energy initially blamed BNFL's intransigence over the spent fuel contracts for its financial crisis, but the nuclear generator's problems were much wider.

The high cost and relative inflexibility of nuclear generation and the failure of British Energy's strategy in the tough UK market meant that even if the waste problem were solved the company would still not be financially secure.

In September 2002, a year and a half after NETA started, the wholesale electricity price in the UK currently stood at £16/MWh – too low for any generator to make a profit. Even those who also had a retail arm were finding it hard, mothballing plants. But British Energy's attempts to access the retail market had failed, leaving it with a limited number of large users and the wholesale market to rely on. And the New Electricity Trading Arrangements (NETA), unlike the previous electricity pool, rewarded suppliers who could offer flexibility – impossible for a company running more than 10 GWe of nuclear plants designed for full power baseload operation.

The pool called on generators to supply every half-hour and paid them all at the price of the highest bidder. That allowed British Energy to offer electricity at zero and be sure it would be called on. But under NETA, suppliers and buyers make bilateral contracts and either side could be penalised if at the time of dispatch they are 'out of balance' – using or supplying more or less than their contract. NETA drove down wholesale prices and left British Energy with nowhere to hide. The drop was estimated to have reduced British Energy's pre-tax profit by some £225 million in FY 2001.

British Energy argued that it was not treated fairly. For example, it said nuclear electricity should not attract the climate change levy, which added 0.43p/kWh to the price of most electricity for non-domestic users and was intended to convince companies to switch to CO_2-free sources. BE argued that since nuclear energy does not emit CO_2 it should qualify as exempt – especially since so-called good quality CHP schemes, some of them gas-fired, could be exempted. BE estimated that the change would be worth £80 million annually. The company also complained about distortions in local taxation. But these were small sums set against BE's losses.

A deal under which British Energy would operate BNFL's Magnox stations for an annual fee, which had been under discussion, would also not be large enough to put British Energy back into the black. What really put it, according to then chairman Robin Jeffrey, 'between a rock and a hard place' was the cost of dealing with its waste. He argued that despite the UK's low electricity prices British Energy's UK business would make a profit, if its spent fuel were managed under the US system.

Although the pressures on British Energy were well known, and reflected in its falling share price, its insolvency warning came as a surprise. BNFL's Norman Askew had been fairly upbeat when talking about its relationship with BE a few days beforehand. Robin Jeffrey had spoken so positively about the company's finance over the summer months, and particularly at an August city briefing, that it attracted the attention of the Financial Services Authority, although the authority did not find that British Energy misled investors.

Meanwhile, the failure of its UK arm caused concern among British Energy's partners overseas. Its US arm, AmerGen, owned jointly with Exelon, had been a profit earner, bringing in £83 million. Now, however, British Energy confirmed that it was 'in the preliminary stages of exploring the possibility of a sale of its interest' in the company. In Canada, too, British Energy's majority-owned subsidiary Bruce Power 'performed ahead of expectations' in the previous fiscal year, according to British Energy's annual report, and work towards the restart of Bruce A was said to be ahead of schedule.

British Energy approached the government in early September 2002 seeking immediate financial support and discussions about longer-term restructuring. It withdrew from the stock exchange and halted trading in its shares, which had fallen from over £2 at their peak to just a couple of pence in value.

Citing nuclear safety and the security of electricity supplies, the government provided 'short-term financial support' in the form of a credit facility for up to £410 million in respect of the company's working capital requirements and cash collateral to support its trading activities in the UK and North America. The facility was increased to up to £650 million on 26th September and extended for two months in order to give sufficient time to clarify the company's full financial position and to come to a clear view on the options for restructuring the company. Additionally, the government took security for the credit facility over British Energy's assets 'in order to give protection for taxpayers' money'. The government also began discussions over British Energy's long-term future.

The support was in danger of falling foul of EU rules against 'state aid' being provided to companies or industries. But on 27 November, the European Commission announced its approval on the provision of rescue aid to British Energy on the basis that a restructuring plan, a liquidation plan, or proof that the loan had been reimbursed in full would be provided and communicated to the Commission within six months.

British Energy announced a proposed restructuring plan at the end of November. On the same day, the government set out the limits of what it was willing to do to support a solvent restructuring. It said it would play its part by contributing significantly to British Energy's historic spent fuel liabilities, underwriting the new arrangements announced by the company to fund decommissioning and other nuclear liabilities and continuing to fund the company's operations while the plan was agreed and implemented.

The government also said it would bring forward legislation to enable it to carry forward its part of the proposed restructuring or, if the restructuring failed, to acquire BE or its assets. The Electricity (Miscellaneous Provisions) Bill completed its passage through Parliament and received Royal Assent on 8 May 2003.

On 14 February 2003, British Energy announced that it had reached agreement in principle with its financial creditors on its restructuring proposal. On the basis of that announcement the government confirmed that it was willing to extend the Company's credit facility at a reduced level beyond 9 March, and was prepared to seek state aid approval from the European Commission for the company's restructuring plan.

On 7 March, the Secretary of State announced that the state aids submission was being sent to the European Commission that day; that BE had repaid to government all outstanding amounts under the credit facility; and on a contingency basis, it would continue the credit facility to BE, with the maximum amount available being reduced from £650 million to £200 million until either 30 September 2004 or the date on which the restructuring plan became effective.

6.3.11 2002–2003: summarising the year

The extent of the debacle was made clear in June 2003 when British Energy published its annual report for the year gone by; it reported a £4.29 billion loss for the year to end March. Total exceptional items – which included write-downs of £3.587 billion ($5.954 billion) for nuclear assets and £151 million ($251 million) in the value of its Eggborough coal-fired plant – came to £4.162 billion ($6.909 billion). Other exceptional items included write-downs in its nuclear decommissioning fund, the AmerGen decommissioning fund and other costs associated with the restructuring.

Before exceptional items, the group had made a loss of £274 million in the UK, compared with a £42 million profit the previous year, somewhat alleviated by a profit of £144 million from the company's North American activities, including a £97 million contribution from Bruce Power in Canada, which by then had been sold as part of the rescue deal.

Total UK nuclear output fell to 63.8 TWh, compared with 67.6 TWh in the previous year. The gas circulator problems at Torness were cited as the major setback, resulting in around 4 TWh in lost output. Technical problems were also experienced at Dungeness B and Heysham 2. The 2,000 MWe Eggborough coal-fired plant generated 5.7 TWh.

UK production costs (including Eggborough and corporate overheads) had been £21.70/MWh compared with £20.30/MWh (3.37¢/kWh) in the previous year, while its achieved price had fallen from £20.40/MWh (3.39¢/kWh) to £18.30/MWh (3.04¢/kWh).

According to Gordon MacKerron of NERA Economic Consulting, the UK government chose to rescue British Energy for two main reasons. First, it accounted for over 20 per cent of all UK generation, and rapid closure of this capacity would have run serious supply security risks and have caused wholesale prices to rocket. And second, it would have taken 18–24 months for a potential new owner of a nuclear station to obtain a site licence and the costs of acquiring a licence are significant.

The rescue package was:

- UK government makes £150–200 million per year direct subsidy for 10 years,
- UK government orders £120 million discount off BNFL reprocessing deal,
- UK government underwrites all long-term decommission and waste liabilities,
- Equity and debts written off, and new lower-value shares and bonds issued,
- Shake-out in senior management: new chairman and CEO appointed,
- BE in future would make 'sweep' payments from any profits to a replacement liabilities fund,
- Requirement to obtain approval under EU competition law.

British Energy was also required to commit as much as possible of its output in long-term contracts – a decision that protected it from the falling spot market but that proved less successful in 2004 as the market tightened and prices began to rise. By 2005 BE was selling around 60 per cent of its production direct to the UK's very large users – bypassing both the distribution networks and the suppliers – and it aimed to continue increasing that proportion.

By 2005 British Energy was beginning to recover. The new chief executive Bill Coley told *Utility Week* 'Our unplanned reactor trip rate is now the lowest in history; lost time accidents are the lowest in history, including contractors. We've worked off 55 per cent of outstanding outage defects.' The company could also point to some long-term positives like the extension of the Dungeness B reactor lifetime by ten years, and further potential extensions down the line. The most effective change, however, was a huge increase in gas and oil prices in 2004 and 2005 and a commensurate increase in power prices. That put British Energy back in the black on its operating account.

What is more, while a 2003 Energy White Paper was sanguine about the possibility that the UK would rely heavily on imported gas for up to 70 per cent of its power, with a heavily subsidised contribution from renewables that would reach 20 per cent by 2020, the rocketing price of fuel and electricity sent politicians and the energy industry back to their calculators. Much of the UK's coal capacity would shut down at the same time as the AGRs would reach the end of their lives. What would fill the gap? And, with the news that the Kyoto Agreement to reduce carbon dioxide emissions would be extended past 2012, how could it be replaced by a carbon-free source? It was no longer clear that the UK's electricity supply was guaranteed in the coming decades: in 2005 the government announced another energy review that included potential nuclear new-build among the future options.

Panel 6.6 Parliamentary debate – British Energy

A bill that would allow the UK government to take over British Energy was passed by 314 to 167 votes in its second reading on 27 January 2003. In a four-hour debate, members of parliament discussed the reasons for the nuclear company's failure and the future of nuclear power in the UK. Highlighted here are some points made by the MPs.

> Energy Minister Brian Wilson: I know that there are some ... who would want all those nuclear stations to be shut down tomorrow. I strongly disagree with that view. For one thing, the sudden shutdown of a nuclear reactor is not a practical option. ... Continuing to run [the stations] generates surplus revenue, which can then be put towards paying for the liabilities that are already incurred. ... The government has agreed to contribute significantly to British Energy's historic nuclear liabilities. However, British Energy will in turn pay its own way in future. That is an important principle: liabilities that arise from future operations at its stations must be paid for by the company.

Conservative energy spokesman Crispin Blunt: [The nominal nuclear capacity, at 22%, is on a par with the UK's excess capacity.] If the nuclear stations were shut down there would still be a margin, provided by a combination of power supply through the interconnector to France, demand management and the 10% margin available by lowering voltage at peak demand. Prices would also rise, but supply would continue. We could switch off all British Energy's contribution to the grid and, on current figures, it could still supply the market on the coldest day of the year. ... If the company had gone into administration, the government would have been left with the liabilities that they have already accepted explicitly... However, the administrator would be able to refloat the company on the basis of the cash-generating assets, which have a significant positive value ... We would have had a nuclear generator ... fully in the private sector, with an opportunity for a transparent and sustainable method of internalising the costs of its environmental pollution.

Russell Brown (Labour): Can he tell the House which third party in the wide world would be interested in taking over the running of those sites?

Martin O'Neill (Labour), chairman of the Select Committee on Trade & Industry, which has reported on British Energy and UK energy: While British Energy was technically sound ... it was at best commercially unlucky and at worst downright incompetent. ... If British Energy had had flexibility and had become a broad-based energy company instead of being almost entirely a generator, we might not have had this debate. ... The reason above all else why little concern was given to the possibility of financial failure is that it takes a particular sort of incompetence not only for a company to run a utility at a loss, but for it to fail to recognise that it is working in a rigged market... [Until 2001, when NETA was introduced, British Energy could bid into the 'pool' system and would be paid for its supply at the price of the most expensive capacity called on.] If British Energy had been as commercially aware as it should have been, it would have acted immediately. Other generators did that and were able to deal with matters. ... When Robin Jeffrey returned from America ... his ambition was to generate even more nuclear energy. He wanted either replacement or additional build. Frankly, he took his eye off the ball.

Vincent Cable, Liberal Democrat spokesman on trade and industry: Can we have an independent financial assessment? Is administration really more expensive than solvent restructuring? Is keeping production going cheaper than stopping it? ... There is ... a serious problem in that the government is favouring one company over others through the Bill ... It is important to have some sense of the damage being done to British Energy's competitors ... One small part of the market is combined heat and power, which has lost about 75% of capacity ... CHP is receiving no compensation or help ... The government will have to explain to the European Commission why British Energy is being helped while other operators are not.

Paddy Tipping (Labour): As a reflection of how quickly the energy market is changing, let me take the minister back to the performance and innovation unit energy review published last February, less than 12 months ago, in which it said: 'Because nuclear is a mature technology within a well established global industry, there is no current case for further government support'. Things have changed fairly radically in the space of a few months and I predict that they will change radically again. ... As things stand, there is no future for nuclear energy. The private sector has never replaced nuclear with nuclear and will not do so until the distant future when we have resolved the issues about decommissioning, disposal and liabilities. There is a strong argument for building a renewables base while nuclear declines.

Chapter 7

A new nuclear programme for the UK?

Post-privatisation investment in the UK electricity generation industry has been in several distinct phases. Once the final form of the industry settled down in the 1990s the new generating companies invested first in gas-fired generation. Previously, large-scale gas generation had not been an option for the CEGB, gas being considered too expensive to use for generating electricity. But for the new generating companies gas-fired stations were the ideal. They were relatively fast to build, capacity could be added in smaller plants of two or three hundred megawatts and it was fairly flexible in the market.

This so-called dash for gas also had a very useful side benefit, as it generally replaced, or was used in preference to, the country's oldest and worst-performing coal-fired stations. The result was seen over the 1990s in the UK's carbon dioxide emissions, which are much lower per megawatt hour when using gas-fired generation than when using coal. The result was a significant reduction in the UK's carbon dioxide emissions – a purely incidental reduction that we are still relying on to meet commitments under the Kyoto Protocol, as otherwise emissions have been increasing.

Aside from the gas stations, the tendency for the new post-privatisation companies was to 'sweat' existing assets: operate them at peak load and maintain and refurbish them not just to achieve their expected lifetime and output but to surpass it. It meant a huge improvement in the system's efficiency, but it also meant that the UK system had a comfortable excess of capacity for many years, which kept electricity prices very low. That was beneficial for customers and industry, who had low electricity bills, but it was unlikely to give power companies the market signal (in the form of rising prices) that would prompt investment in new generating capacity. In fact, in the late 1990s generating companies were 'mothballing' their least efficient plant: placing them in long-term shutdown until the price rose sufficiently to make it worth operating them again.

This phase of sweating conventional assets lasted well after 2001, when the New Electricity Trading Arrangements sent prices still lower and left many generators unable to operate even their most efficient plant and get an economic return. However, it also overlapped with the next phase of large-scale investment: a government-imposed requirement to invest in wind generation (known as the 'dash for wind') and other renewables.

The new phase was prompted by the growing realisation that the worldwide production of carbon dioxide from burning fossil fuels was altering the composition of gases in the atmosphere and causing climate change that could have catastrophic effects on human activities worldwide. The consensus was that the effects would be dramatic, from changing weather patterns to increased sea levels. An overriding requirement for energy policy – and other activities – was to reduce the amount of carbon dioxide emitted into the atmosphere.

The UK government, with many others, committed to reducing the country's emissions of greenhouse gases (mainly carbon dioxide, but including methane and others in a so-called 'basket' of greenhouse gases) to below 1990 levels by 2012 in a commitment signed at Kyoto in Japan. The burden would fall on other sectors as well, but the power generation industry, as a major emitter, would be expected to make a large contribution to the emissions reduction.

The 'dash for gas' had in fact initially put the UK in a strong position to achieve its target when it replaced older coal stations in the 1990s. However, that one-off change was, by 2000, being chipped away by increasing electricity use and by rising gas prices that made it more financially attractive for electricity generating companies to use more coal-fired generation in the mix – in direct opposition to the intention to switch to less CO_2-emitting sources to meet the Kyoto targets.

The government's solution to the need to reduce carbon dioxide emissions was to force generating companies to use renewables for a significant proportion of their generation. The government had tinkered with renewables in the past, auctioning long-term power contracts for renewables projects and reserving part of the non-fossil fuel obligation for renewables projects. From 2001, however, the scale of renewables investment was to be ramped up. The mechanism chosen was known as the Renewables Obligation.

7.1 Renewables Obligation

The Renewables Obligation required electricity supply companies, who sell power to end users, to source a growing percentage of the power they buy from a renewable supply and to prove that they had done so or pay a fine known as the 'buyout'. The renewables proportion required would increase each year from the first year of operation, in 2003–2004, when it stood at 2.4 per cent, with the aim of reaching 20 per cent by 2020. Targets were set immediately that gradually increased the requirement to 10.4 per cent by 2010, and the Obligation was extended a couple of years later to 15 per cent by 2015. The Obligation was made law in April 2002 and was guaranteed to operate until 2027.

Supply companies would prove that they had met their obligation by a system of Renewables Obligation Certificates (ROCs). These electronic certificates were generated each time a megawatt hour of electricity was generated by a qualifying and certified renewable energy generator. The ROCs could be sold with the electricity, or they could be separately traded.

This meant the supply companies had a number of ways to meet their obligation. They could build and operate their own renewable energy generating facilities; they could buy electricity from other renewable generators, along with its certificates; or if other generators had excess certificates to sell it could buy them separately from the electricity. Finally, the supply company could pay the buy-out fee in place of some or all of the certificates.

Buy-out fees would be recycled pro rata to companies that had produced ROCs. This increased the value of certificates, as the Obligation was extremely challenging and it was fairly certain that there would be a significant buy-out fund.

The Obligation was intended to bring near-market renewables into operation quickly and at large scale. In the UK, this initially meant the Obligation was met by existing small hydro projects and by burning the gas produced at waste landfill projects. After a few years new projects would begin to come on stream and it was clear this would mostly be wind power, as several European countries, notably Germany and Denmark, had been heavily subsidising wind power in recent years and as a result prices had come down to near-market values.

There were other renewable options but their contribution was likely to be much lower. They were:

- Hydropower. Most of the UK's potential hydro sites had already been exploited so large gains were unlikely.
- Photovoltaics. This had also been supported in Europe with large-scale programmes like Germany's '100,000 roofs'. However, it was still very expensive and even though the European programmes had been successful, the total power generated was insignificant when compared with the country's major sources.
- Wave and tidal generation. This was at an early stage of development and thought unlikely to make a significant contribution before 2012.
- Biomass-fuelled conventional generation. Developing biomass was considered to be hampered by the need to develop fuel sources and a fuel supply route, as well as adapt the generating technology. As such it was unlikely to make major contributions by 2012.

The Obligation successfully prompted an increase in renewables. By 2005, when the Obligation was 5 per cent, there were 113 wind farms in operation onshore and two more offshore, while a further 22 were under construction and 77 more had planning permission. Those 212 wind farms together had a capacity of 4,100 MW.

7.2 The gap ahead

The Renewables Obligation is intended to stimulate investment in renewables. What it does not address is the need to replace existing power-generating capacity – and to replace it with new sources that will not increase, or better still that will decrease, carbon emissions. Electricity consumption is still rising by over 1 per cent annually. Major retirement of capacity is expected over the next 20 years and it is far from clear what will replace it.

The retiring generating stations will total at least 9 GW, out of the UK's total generating capacity of 70 GW. That is around 20 per cent of the whole. This includes the whole of the UK's nuclear fleet, apart from Sizewell B.

The eight Magnox reactors still in operation at four stations (Dungeness A, Sizewell A, Oldbury and Wylfa) are expected to shut down within the decade, taking some 2.5 GW out of the system. They will be followed into retirement over the next 20 years by all the AGR fleet, which are currently planned to shut down between 2015 and 2025. There may be some flexibility in this shutdown schedule as British Energy is likely to be able to extend the lifetime of some of the AGRs for up to 10 years (as has already happened at Dungeness B) and certainly intends to do so if it can be technically justified. The AGRs cannot hope to extend their lifetime as far as PWRs, which in some cases may operate for 60 years. The AGR shutdown will remove 9 GW of capacity from the grid, and what is more it will remove capacity that is an important component of the UK's carbon-free generation.

On the same timescale, several of the UK's largest coal-fired power stations will be shut down. This is because they will be in breach of the EU's Large Combustion Plant Directive. This regulation from the European Union aims to reduce emissions of several pollutants – sulphur dioxide, nitrogen oxides and particulates – from large combustion plants. It does not apply only to combustion plants whose purpose is power generation: refineries, steelworks and other major industrial sites may also be affected. The Directive requires all new plants of this type to include equipment to minimise these emissions. Operators of existing plants, which include the UK's coal-fired power stations, must decide whether it will be economic to install equipment to reduce their emissions of these pollutants. If not, the company must agree to shut the plant down. These plants can operate for only 20,000 h between 2008 (when the decision must be taken) and the final shutdown deadline of 2015.

Generating companies in several cases have opted not to upgrade but to shut down their stations, including some large coal stations such as Longannet in southern Scotland.

The two waves of shutdowns could leave the UK short of around 20 per cent of the generation it requires, and replacing it could make dramatic changes in the country's carbon dioxide emissions. The nuclear stations are largely free of carbon emissions at the point of generation, whereas the coal-fired stations have of course among the largest emissions.

In 2003, when the government last examined the energy policy in an Energy White Paper, it seemed likely that both the nuclear and coal stations would be replaced by more gas-fired stations. This was not unattractive in carbon dioxide terms – swapping low and high emissions sources for one with medium emissions may leave the UK with a broadly equivalent carbon dioxide burden or even with a slight benefit. But it would mean that up to 70 per cent of the UK's electric power could rely on gas, and imported gas at that, as the UK's own gas resources in the North Sea are already beginning to reach the end of their lives. In fact, in 2005 the UK became a net importer of gas for the first time.

The possibility that the UK's generation needs would be met by imported gas was of concern. Many other countries are also turning to gas-fired generation, while gas resources are limited and geographically restricted, and it was suggested the country would be last in the queue for supplies from Russia and the Middle East.

However, in its 2003 Energy White Paper the government decided that this was not unacceptable. It concluded that expansion of import routes for gas fuel into the UK was required and that terminals to import liquefied natural gas (LNG) should be built (previously the UK had no significant LNG infrastructure). To this should be added an upgraded pipeline capacity to ensure that sufficient supplies could be imported and that there would be enough competition among suppliers.

The White Paper also proposed a raft of new measures to improve energy efficiency. It argued that energy use and economic growth had been decoupled in the 1990s and that there was much more that could be achieved in energy efficiency. A successful efficiency programme could cut or even reverse the growth in energy and electricity consumption. If that happened then the growing renewables capacity and additional sources of gas would together be enough to meet the UK's requirements. What was more, if there were no energy gap looming in the 2020s, there would be no need for replacement of nuclear or coal stations.

As a result, the Energy White Paper said that the government had decided that new nuclear stations would not be required at this stage and placed them well behind renewables, energy efficiency and gas as the favoured way of meeting the UK's power needs.

However, in late 2005, just two years later, the government found itself setting up a new energy review: one that would examine whether the UK should invest in new nuclear stations and carbon capture and storage. What had happened?

7.3 A different paradigm

By 2005 the days of overcapacity were gone. Over the past two winters generators had brought back into operation most of their 'mothballed' plants to meet peak demand, and in response to consequently much higher electricity prices. The years since 2000, and especially since 2003–2004, also saw huge increases in gas prices, which were passed on from generators to suppliers and end users, amid fresh concerns that the UK would run short of gas supplies.

It was clear that the pressure on day-to-day gas supplies would ease over the next few years, as the new LNG terminals and the increased pipeline capacity promised in the White Paper began to come online. But the larger questions over gas supply would be likely to remain. Were UK companies in strong enough competitive positions to secure sufficient supplies of gas in competition with companies in other countries? The UK's weak position seemed to be confirmed by an EU investigation that found anticompetitive practices in the European gas market. The UK's physical position seemed equally insecure, despite the new terminals and increased pipeline capacity.

Meanwhile, the worldwide pressure on gas supplies had contributed to volatility in gas prices, as had a series of disasters such as hurricane Katrina, which laid waste parts of the southern USA and knocked out oil and gas refineries in the process, revealing just how vulnerable fossil fuel supplies could be. This was underscored at the end of 2005 when a gas price dispute between Russia and Ukraine caused concern across Europe. Russia's gas pipelines traverse – and supply – Ukraine. When Russia tried to impose large price increases on Ukraine it threatened to cut off supplies. That placed other European supplies in danger, and Ukraine's response that it would take what it needed also left utilities afraid that remaining supplies to western Europe would be inadequate.

Gas prices rose and fell by several hundred per cent over the year. For electricity generators, expensive or uncertain gas supplies meant that it was economic to switch to using coal-fired generation in preference to gas-fired, with consequent increases in carbon dioxide emissions.

Meanwhile, the energy efficiency measures planned in the Energy White Paper, although partly implemented, had not achieved the type of deep cuts in carbon dioxide emissions or energy use that would enable energy supply companies to put off investment in new plants.

The long-term situation looked no more certain: gas prices might fall, but it was unlikely they would be as low as they had been at the end of the 1990s and they might just as easily rise again. Even if they fell, they would always be volatile. Energy consumption might fall if energy efficiency measures were successful but in 2005 the trend was in the opposite direction, as it had been in every year since the White Paper.

Energy use was still increasing, the trend towards gas was making energy prices more volatile and expensive and capacity margins were becoming slimmer. It seemed more than possible that the energy gap that had been feared when the coal and nuclear stations closed would be a reality and the White Paper solutions would not fill it.

As a result, in 2005 the government announced a new energy review, which would look at what measures would be needed by 2020 and beyond to tackle climate change and to ensure secure and affordable energy supplies in the UK. The review would look at:

- Ensuring emissions reductions,
- Attaining reliable energy supplies,
- The role of nuclear power,
- The potential of carbon abatement and low-carbon technology,
- Achieving affordable and adequately heated homes.

The role of nuclear power in the UK's future energy mix, and what actions from government would be required to allow energy companies to make the necessary investment if it was required, should form a key part of the completed report.

7.4 Is new capacity required?

It should be said that when the review was announced many disagreed that there would necessarily be an energy gap in the future.

Scottish and Southern Energy (SSE) was typical of energy companies that believed the gap would be real – even if there were major successes in improving efficiency. The company said in evidence to the Commons Environmental Audit Committee that if the UK's nuclear and coal-fired power stations were phased out, as expected, around 2020, it would leave the country with capacity to generate around 200 TWh of electricity each year. The company said that if electricity demand increased, as expected, by 1.5 per cent annually it would reach 400 TWh by 2020; if electricity demand fell by 1.5 per cent per year the UK would still be consuming 280 TWh in 2020. The shortfall would therefore be between 80 and 200 TWh per year.

In response, Brenda Boardman and Catherine Mitchell of Warwick Business School defended the energy efficiency policies laid out in the 2003 Energy White Paper. They told the same committee that it was far too early to say that the White Paper's efficiency targets could not be met and that 'if there were appropriate policies for low carbon generation and energy efficiency in place, a generation gap would not occur'. Dr Mitchell said there was enormous potential for energy efficiency and microgeneration, and the new energy review 'should be about how a robust plan is put in place to make sure that the fundamental recommendations of the White Paper are delivered'.

WWF-UK also said it disagreed that energy demand would continue to rise 'indefinitely and unavoidably'. It said the energy gap 'is essentially a political choice', citing work for the 2003 White Paper on energy by the Performance and Innovation Unit, which estimated energy efficiency could reduce UK energy demand by 31 per cent.

Meanwhile, the Imperial College's energy policy unit estimated that the renewables potential in 2025 could be 230 TWh per year – potentially, enough to fill the gap.

Whatever the potential for energy efficiency, electricity companies were working on the basis that the closure of most of the existing nuclear stations – possibly by 2023, although some extension may be possible – and several coal stations will require new capacity to be built.

7.5 What type of capacity?

If extensive new capacity is to be needed, what form should it take? In one sense, in the UK, that is a decision for the power generating companies, who can build whichever type of power station they think is appropriate to their needs. The privatised electricity system is, after all, a 'market' system where no organisation or function is given the responsibility of ensuring 'security of supply'. The market is expected to provide the incentive for generators to make their own investment and to choose the most economic solution. But all forms of generation have different characteristics – technical and economic – and more factors come into play.

All generation options have strengths and weaknesses that alter their value not just for the generating company but for the electricity grid as a whole.

Very briefly, the strengths and weaknesses of various forms of generation are as follows.

Gas turbines can be started up and shut down, if necessary, over a period of hours to less than an hour, so they can be brought on line to meet peak loads. They are not very flexible in operation (although more recent versions are being designed to operate more economically at part-load) but they can be sized at between one and several hundred megawatts, so they can allow mid-scale additions or removals of capacity from the system. They require constant fuel feed-in during operation. 'Combined cycle' gas turbines include a steam turbine that means their overall efficiency is higher, but their flexibility is lower.

Coal is a relatively flexible form of generation and most plants can operate at part-load. They can be cycled up and down from hour to hour to meet changing demand. Fuel is fed in constantly during the operation. An important long-term issue is that their fuel can be stockpiled so there is a reserve in case of need. Both coal and gas generation includes large rotating machinery (the turbine) that produces the electricity; in the context of the grid, this means they add stability to the operation. That is because most power plants are set up to detach automatically from the grid if there are large disturbances in the grid supply as a self-protection measure. Disconnection, in turn, creates a new disturbance so faults can propagate and the effect can spread. The mechanical generators in coal and gas plants to varying degrees have momentum that will carry them through grid disturbances (this is known as 'fault ride-through') and this provides stability to the grid as a whole.

Hydropower is among the fastest responses in the system. There is no fuel to burn; so as long as there is water in the associated reservoir or river it is only a matter of opening the gates within the plant, so water passes through the turbines, and generation is available within seconds to minutes. This is the attribute employed by pumped storage plants: water is pumped uphill to a reservoir at times when there is excess power available on the grid and released to generate at peak times. However, from a 'fuel' point of view, over the year there are periods when water levels are low and this can force so-called 'run of river' plants out of operation. Operators of hydro plants with water stored in reservoirs have to decide whether to use their stored water to generate now or save it for a later date when it may be needed more. This is a mainly financial decision in a mixed system like the UK's but far more important in countries such as Norway that have a very high reliance on hydro power. Elsewhere it has led to accusations of hydro companies 'gaming' the market – holding back water supplies unnecessarily to exacerbate a power shortage and force up the price of electricity.

Wind power, of course, is only available when the wind is blowing. That means that it is impossible to guarantee that power is available when it is most needed (at peak times, for example). The power has to be accepted onto the grid whenever the wind blows and other forms of generation have to be cycled up or down to adapt. In practice, in the UK this is hardly an issue, as wind power penetration is very low and wind forecasting is very good, not least because in the current market decisions on how much power is available and required are calculated up to an hour in advance, and over such timetables short-term prediction is extremely reliable. In the UK, National Grid has said that the variability of up to 10 per cent wind is no larger than the general variability within the system – that is, it gets 'lost in the noise' and the system can handle, albeit at increasing cost, much higher levels. What is more, wind farms can offer fast response in some circumstances: if a large wind farm is in operation and expected to be so for the next hour, it can provide extremely fast response to changes in demand on the grid over the short term. In that case the wind farm would be gradually turned down in advance of an expected peak and then turned up quickly when needed, then kept there as slower-response forms of generation like coal stations are brought up to power. But predicting whether the wind farm will operate over days, weeks or months is progressively less reliable. Recent work has confirmed that there is almost never a situation when there is no wind blowing anywhere in the UK, but there are frequent periods when smaller regions have no wind. As a result there is a limit to how much conventional generation they can replace on the system, and although estimates vary depending on other grid characteristics its value may begin to decrease above 10 to 20 per cent wind. Grid stability may also be affected at high wind penetration because wind is generally designed to disconnect in the event of a fault, although the 'fault ride-through' of conventional generation can now be replicated using electronic systems.

Nuclear stations are likely to be among the largest on the grid – Sizewell B, for example, provides 1,400 MW – so they provide an enormous input of power. What is more, they can provide that power over a long period as fuel loading is infrequent, even when done off-load, and they can operate for one or two years between shutdowns depending on the operating regime. But they are extremely inflexible in operation. Although it is possible in some cases to vary their output slightly, it is technically and economically undesirable (although this also means they have good fault ride-through characteristics). What is more, they are the slowest option to bring in and out of operation as they can take days to bring up to full power. An unexpected shutdown (as can happen to all types of plant for various reasons) takes a large power input out of the system very suddenly.

The UK's power system is well placed to cope with all these different sources, and indeed their diversity gives system operators a useful set of different options to meet the system's varying needs. Along with, and sometimes resulting from, the different sources' system characteristics are their very different economic pressures. As well as assuming that all these power sources are available from large power stations, the opportunities for 'microgeneration' are also increasing. These are small projects sized at a household or local level from just a kilowatt to a few megawatts. While a variety of different designs are entering or soon to enter the market, the UK's local electricity systems are not well designed to accept them. So while they have plenty of potential it seems unlikely that they will meet a significant proportion of the UK's needs over the next two decades, when decisions on whether major new capacity is required must be taken.

7.6 The economics of operation

As well as grid management issues, different power generating options are under different financial pressures.

Nuclear stations represent an investment extreme, in that capital costs for the project are very high in comparison to running costs. The high price of building nuclear stations means that much of their ongoing cost is in fact interest payments on the construction expenditure. Fuel and running costs are a relatively small part of their operating budget. Hydro and wind plants have the same situation of high upfront costs and their fuel costs are, of course, zero.

In comparison, gas turbines are relatively cheap to build. That means that interest charges over the plant's lifetime are low, but the cost of fuel is high and may change unexpectedly, as in the UK gas market over the 2003–2005 period. This may change plant operating decisions in later years. With a much lower interest burden, a gas turbine operator has the option of halting generation from its plant if it is a good financial option or of reserving its gas fuel and providing power only at peak times when demand – and therefore price – is highest.

But because interest payments are fixed (within variations in interest rate) for nuclear stations much of their operating costs are inescapable whether the plant is operating or not. A nuclear plant operator must maintain income from the plant.

In the CEGB and similar monolithic state generators this may not be a problem. The UK demand, for example, varies between over 51 GW at peak times (during a winter's day when industry and domestic consumers are drawing most power) and 30 GW on a summer night when demand plunges. A large power company operating a mixed fleet of nuclear and other forms of generation can meet this demand by using its nuclear stations as a 'baseload' supply – that is, operating continuously to supply some or all of the constant 30 GW load. It can then use more flexible forms of generation to match varying demand above the baseload.

This can become a real problem if a country is over-committed to nuclear power: France's 58 nuclear stations may be forced to vary their output as the country's load changes, which as we have seen is undesirable. It may also be problematic for companies with a smaller demand to satisfy as in the UK, where demand now is split between several supply companies. Initially, the 'pool' took care of this issue (see Chapter 6): nuclear could bid in at 'zero' price and be guaranteed to run. Now British Energy deals with the issue by selling as much as possible of its output on long-term contracts to large users, whose demand is more stable. This reduces British Energy's exposure to the day-to-day variations of the consumer market.

Increasing interconnections between European power grids may alleviate this problem. For example, rather than vary the load on its nuclear stations, France can offer excess power to other countries (in addition to the major power exports it is already making to its neighbours). The UK's current interconnections, however, are strictly limited. There is a 400 MW interconnector with Ireland and a 2,000 MW interconnector with France.

7.7 Filling the supply gap

The scenarios used in 2003's Energy White Paper and by generating companies for forecasting their likely needs assume a gap of up to 20 per cent when the nuclear and coal stations close down.

It is possible that energy efficiency will reduce this gap; but it is also possible that alternative scenarios will increase it. One example is changes in the fuel used for transport. Oil is the most commonly used source of transport fuel, but emissions targets are already changing public policy here. Hydrogen-fuelled buses and cars are being operated in demonstration projects or are being developed by major manufacturers and hydrogen is frequently suggested as a replacement energy source for transport and many other purposes. Hydrogen, however, is simply a method of storing energy, and it does not come for free: power is required to separate the hydrogen from water or another source and to transport it to the point of use. That means that any shift to using hydrogen fuel will have to be accompanied by additional generating capacity to meet the need to produce hydrogen. Similarly, switching from fossil fuel to electric vehicles also requires a commensurate increase in electricity-generating capacity.

The Tyndall Centre for Climate Change Research, for example, recently modelled a number of scenarios that would meet the UK's target to reduce CO_2 emissions by 2050. The Centre found that our rapidly expanding appetite for airline travel and global trade would divert almost all the allowable carbon dioxide emissions to those sectors. All our other energy needs would have to be met using electricity or hydrogen generated from carbon-free sources that in some scenarios meant a huge revival and expansion in generating capacity.

In more than one scenario, then, new electricity generating capacity will be required to meet our still-considerable power requirements. Wind at proportions greater than 20 per cent would certainly require some backup power and other renewables are almost fully exploited (hydropower) or are at too early a development stage to provide such bulk power. That means generators can choose between gas, coal or nuclear power to fill the gap.

The arguments for and against gas have been well rehearsed since 2003 and played out in the UK industry: convenience and low capital cost are balanced against price volatility and concerns over supply, and so far it seems it is the concerns that will limit companies' investment in gas to make sure it does not make up the bulk of capacity. That leaves coal or nuclear, or some combination, to make up for the likely shortfall.

Coal has certainly fallen from favour since the dangers of carbon dioxide have become such an important consideration during the past decade. But as we have seen, it has some benefits within the power mix: stability on the grid, flexible operation, a variety of fuel suppliers and the ability to stockpile the fuel.

Investing in new coal now is certainly possible: in late 2005, for example, E.ON announced plans to build a new coal-fired station in eastern England. But the problem of carbon dioxide has to be addressed.

One option is to accept the carbon emissions, albeit while reducing them as far as conventionally possible. The UK's emissions of carbon dioxide are capped by agreement with the European Union as part of the Emissions Trading Scheme (ETS), and are likely to be reduced in future years, but the power companies can use their emissions allowances for the new coal station or buy additional allowances. There may be enough emissions allowances available, but their price is impossible to predict at this stage; and in any case it has frequently been argued that the aim of emissions trading is not to allow major emitters to 'buy their way out of trouble' but to allow major reductions to be made as quickly as possible. The arguments over whether this is an acceptable approach may vary depending on whether the emissions credits are being used for coal stations that replace other coal stations or whether they are being used to replace nuclear, which is at least free of carbon emissions.

The alternative approach for a power generator that wants to invest in a new coal-fired plant is to invest in new technology – carbon capture and storage – to ensure the carbon dioxide emissions are not released into the atmosphere.

Neither carbon capture nor storage are unknown technologies. The chemical and oil industries routinely remove carbon dioxide from their waste or process streams, transport, store and use it. Elsewhere, the oil industry has pumped carbon dioxide into undersea reservoirs to maximise the amount of oil that can be recovered from them, and the carbon dioxide is expected to remain as firmly locked in the reservoir as the oil was before recovery began.

But the process is at a very early stage for large-scale capture and storage from power stations: work has started on demonstration projects, and discussions have begun on who would shoulder the long-term liabilities, for example if a carbon dioxide storage site was found to be unsuccessful and the carbon dioxide was released. Installing the necessary infrastructure may also be a big barrier to large-scale use of these techniques. What is clear is that there will be a cost involved in dealing with the carbon dioxide produced by burning of coal that may make it unattractive to generating companies.

With all the alternatives assessed, it seems more than possible that power generating companies in the UK would consider building replacement nuclear power plants.

7.8 The economics of nuclear power

As we have seen, the economics of nuclear power are difficult to support in a privatised market where returns are expected within a few years. It is difficult to support an investment that is expected to make a return over a 40-year lifetime, in a market where forward prices are only available for less than five years ahead and one where prices can vary rapidly from year to year and there is no guarantee that the market will retain its structure over such a long timescale. Investors in the wind sector, in comparison, require certainty that their market will exist for at least 15 years, which is why the Renewables Obligation has been set until 2027. Would nuclear power become attractive to investors with a similar guarantee or are there other barriers?

There are some benefits from such long-term investments, as Gordon MacKerron of the Sussex Energy Group explained in evidence to the National Audit Committee.

He said 'The costs of wind and other renewables technologies are largely fixed and predictable, so they add an element of stability to the portfolio. Even more important, the costs of these technologies are also independent of fossil fuel costs.' This would apply equally to nuclear investment, but 'there are sizeable and nuclear-specific barriers to new nuclear investment in the UK'.

Tony White of Climate Change Capital identified two nuclear-specific risks as the large up-front costs required to secure the appropriate authorisations and the uncertainties of decommissioning and spent fuel management. He said both these barriers would 'need direct intervention by government'.

Two other barriers were common to all large projects: would it be built to time and cost and will revenues provide adequate returns on capital? Tony White pointed out that in the UK's deregulated electricity market, companies had to work on the basis not of the average power price but on generally lower marginal prices with occasional spikes. Long-term contracts might 'smooth' this pattern but the UK had not developed such a market. That meant that while renewables would be built in the UK because they are supported by the Renewables Obligation, 'some changes may be required to the wholesale power market in Britain before developers commit to the construction of any new capacity: be it nuclear, clean coal and perhaps even gas,' Mr Strong said.

Policy analyst Nigel Hawkins addressed this issue. He noted that the returns from nuclear plant are very long term. In assessing future financial returns from a new nuclear plant, calculated from a 40-year discounted cash flow model, two factors are crucial – the cost of capital and the selling price per kilowatt hour.

Assuming that an 80 per cent/20 per cent debt/equity structure is used, it should be feasible to achieve a cost of capital figure of between 7 per cent and 8 per cent; this range is predicated upon HM Treasury providing a similar guarantee to that applicable to Network Rail. The cost of capital figure is highly sensitive, especially over a 40-year period.

For investors, the unit selling price of the output is also critical. If a 3p per kWh price could be achieved, through the signing of long-term power purchasing agreements (PPAs), the financial case for new nuclear-build would be extremely robust. If the price were just 2p per KWh, however, it certainly would not. Whilst the latter figure currently seems a non-starter, it was less than four years ago that NETA was introduced, from which wholesale prices of below 2p per KWh resulted – along with the collapse of the UK's largest nuclear generator, British Energy.

With the cost of capital and unit selling price at 7 per cent and 3p, respectively, new nuclear-build is a real possibility. But if the figures were 9 per cent and 2p, the numbers simply would not stack up.

By using the published overnight cost for potential nuclear plant designs and scaling them to a virtual 1,250 MW size, Hawkins' calculation yields an overnight plant cost average of £1.2 billion, just under £1 million per megawatt. This figure includes a 30 per cent first-of-a-kind allowance for the first of the assumed four plants.

In terms of running expenses over a 40-year plant life, the operating and maintenance costs come out at between 0.4p and 0.5p per KWh, whilst the fuel figure – after allowing for higher uranium prices – lies between 0.3p and 0.4p per KWh. The waste and decommissioning costs are around 0.1p, a distinctly modest amount since most of the expenditure is deferred for many decades.

Based on a £1.2 billion 1,250 MW nuclear plant generating for 40 years, a total operating cost of around 1p per KWh and a cost of capital between 7 per cent and 8 per cent, these figures should produce a positive equity return if the unit selling price were a minimum of 2.5p per KWh. At 3p per KWh, the projected equity returns are very substantial.

Another analyst, Dieter Helm, argued that the correct mechanism was not to favour nuclear but low-carbon generation, and he noted that the need to make carbon emissions savings was not well served by the existing energy market. He said the ETS time limit of 2012 was far too close, even for investors considering fast-build gas turbines. The government should have a carbon policy in which it set a figure for the amount of emissions savings it wanted to achieve in the long term (such as to 2050, which is the timescale of the 2003 Energy White Paper). The savings requirement could be auctioned among low-carbon suppliers – including nuclear, energy efficiency and renewables – and the least-cost option would be the biggest winner.

Professor MacKerron told the committee that the most important economic obstacle to nuclear power 'is that no one really knows what a new nuclear plant would cost to build, and it will be impossible to know for some time. Bearing in mind that the construction cost and time of a nuclear plant are the single most important determinant of the economics, this is a serious obstacle'. Hawkins put the two major issues – planning and waste management – at the top of his list of economic and practical barriers.

Figure 7.1 *Low-level radioactive-waste disposal in drums, France [CEA (Commissariat a l'Energie Atomique)]*

EDF operates over 50 reactors in France and is a major electricity generator and supplier in the UK, and its chief executive Pierre Gadonneix said that the company would 'certainly' be interested in investing in reactors in the UK. But chief executive Vincent de Rivaz told the select committee that 'In the absence of risk mitigation by Government, nuclear will not be built in the current market'. He identified licensing as one of the most important risks and Professor MacKerron concurred, saying: 'UK licensing requirements tend to be more onerous and costly than those of other countries. This could be different in future, but again no one would expect an investor to risk much money on it. The conclusion is clear: no single-number estimate of nuclear construction cost in UK conditions has any real meaning.'

Along with the licensing cost, the capital cost depends strongly on whether a single plant or series is built – one reason why EDF quoted nuclear construction costs (calculated by other organisations) varying between £865 and £1,330 per kilowatt. Those costs assumed that 'the vendor's standard design is retained without modification'. This is far from guaranteed: the Nuclear Installations Inspectorate required extensive changes to the design for Sizewell B, although it was based on a standard US design. Without changes, 'development costs are expected to be approximately £250 million for a first of a kind reactor . . . falling to approximately £100 million for subsequent reactors in the same series,' EDF said.

In written evidence EDF said 'Investors in renewables require projects to have off-take contracts for power, ROCs and LECs. . . . Similarly large CCGTs being developed by non-vertically integrated companies typically seek off-take contracts to lower the risk profile. . . . It seems reasonable to assume that similar levels of uncertainty regarding revenue streams, to reduce risk, would be a requirement of financiers before they would lend to new nuclear projects'.

EDF said one option was 'offering operators of new nuclear plant a carbon avoidance contract that would pay operators at a predetermined rate on output achieved (and CO_2 emissions saved). The payments could be funded by revenues raised from the auctioning of CO_2 allowances' from the EU's Emissions Trading Scheme. EDF said 'The retention of a new entrant reserve in its current form in ETS would discriminate against carbon-free technology and support the development of non-carbon free technology'.

The need to build not one reactor but a series of eight or ten introduced further problems: Tom Burke of Imperial College said 'the EU and OFGEM would raise questions about the degree of collaboration that would be necessary between the utilities to, as it were, recreate the CEGB, because in effect they would have to do something close to that to create the purchase of eight to ten nuclear reactors in a guaranteed order of sequence. Imagine the negotiations that would have to go on.' Professor MacKerron agreed: 'Non-nuclear electricity generators would be unlikely to sit back and accept as reasonable the "state aid" that government support of this kind would involve'.

Electricity companies have said that diversity of supply is the ideal and they would need both renewables and nuclear stations in their supply mix, but there is a fear that investment is limited, and major commitment to one would necessarily reduce the opportunities in the other sector. If the government supported a large nuclear programme, MacKerron said, 'the prime effect would be "crowding out" other low-carbon investments, especially renewable energy' and Scottish & Southern Energy pointed out that 'If, at any stage, there appears to be any dilution in the government's commitment to renewable energy, investment is likely to suffer'.

7.9 Types of support

Companies considering whether to invest in new nuclear would no doubt welcome a nuclear obligation, or low carbon generation fund. Other forms of support could include a cap on construction costs and a tax credit for the first few years of generation – both subsidies offered to new-design nuclear projects in the USA, under the 2005 Energy Bill. However, as EDF chief executive Vincent de Rivaz made clear it is the uncertainties in the UK licensing system that present perhaps the easiest targets for government support.

The UK's regulatory system is 'not prescriptive'. It does not lay down pipe-by-pipe requirements for the reactor but instead requires the operator to demonstrate its safety to the NSD's satisfaction. This should enable the UK to encourage innovation and take on new best-practice quickly, but to a company considering whether to license a reactor in the UK it represents a huge uncertainty. Sizewell B, for example, was based on a very well-known design, the so-called Standard Nuclear Unit Power Plant System – Snupps for short – from Westinghouse of the US. Initial design work for Sizewell B-type reactors was completed back in the 1970s and many of the 150 PWRs already operating around the world were using the Snupps design.

The design was licensed for the UK by the Nuclear Safety Division of the Health and Safety Inspectorate (and its Nuclear Installations Inspectorate). Despite the familiarity of the design overseas, UK licensing was not just a rubber-stamp exercise – one estimate suggests that translating Snupps into the Sizewell design added 13 per cent more pipework, 22 per cent more cabling and 75 per cent more structural concrete. Since the UK's regulatory philosophy is very different from that of the USA, simply rewriting safety cases and supporting documentation was a mammoth task.

New designs claim to be simpler – Westinghouse's new PWR, the AP1000, for example, claims 45 per cent less seismic building volume, 50 per cent fewer valves, 35 per cent fewer pumps, 80 per cent less pipe and 85 per cent less cable than older designs. As a consequence they claim to be cheaper, more reliable, safer and create less waste during operation and at the end of their lives. But at the start, that means a new UK licensing process.

Once the design has been licensed, and with the general support of the government for a nuclear programme, you might be forgiven for thinking it is time to start building, but there are still areas of uncertainty. You need a site and planning permission. And while existing nuclear sites may be ideal locations for replacement stations that is no guarantee that the planning process will be fast or unobstructed. The Sizewell B inquiry took over two years despite the fact that Magnox stations had been operating at the site for two decades.

Finally, approval from the Environment Agency is needed before the nuclear regulator can approve the start of construction. After completion of the plant the EA must again approve the environmental impact of the project before it can be operated, as it is the body that provides authorisation to release the small quantities of radioactivity required in routine operation.

All these licensing and planning requirements cause delay in the early stages of reactor construction, which as we have seen are particularly important in the nuclear case because of its very high capital cost.

These are the obstacles referred to by de Rivaz and others. They would like more certainty from the NSD about what will be required in the licensing process. The stumbling block is that the NSD has no staff available at this stage to carry out a licensing effort likely to stretch over several years. On the contrary, its regular inspection regime and the reorganisation required by ongoing changes in the nuclear industry 'will stretch our resources', the NSD said in its annual plan. This is a long-standing problem: in 2001, British Energy and BNFL asked the regulator to start a pre-licensing review for the AP1000. The regulator declined because of a lack of adequate staffing, according to a government briefing.

New research programmes – including a £5 million fund for nuclear R&D from the Engineering and Physical Science Research Council – are intended to aid recruitment in the sector. Nevertheless, the shortage of well-qualified staff is likely to continue, as the NSD acknowledged in its long-term strategy. It said it will need to recruit regularly in the foreseeable future, not only to maintain full capacity but also to cater for increasing retirements and resignations. About 30 per cent of NSD's inspectors are due to retire over the next six years and the pool of potential applicants is becoming more limited as the industry continues to shrink. A new-build programme would doubtless not only bring new recruits into the industry but could also call on consultants and staff from overseas. Nevertheless, building the regulatory muscle to license one or more new designs will require considerable planning and investment.

Any potential investor in a UK nuclear programme would like to see the scope of issues that can be brought into the planning inquiry to be reduced. At the moment, and following the example of Sizewell, issues like the need for nuclear power and the design of the station can be debated again at the planning hearing. An approach that covers local planning issues only would be far quicker and could be accomplished without new legislation.

7.10 What are the choices?

If the government did provide the right sort of environment for power companies to invest in new nuclear plants, who might own and operate them? British Energy is a specialist, but its experience of light-water reactors is limited to Sizewell B, and in any case it has so far set its face against investing in new stations.

However, British Energy is not the only option. As stated above, EDF Energy is both an important power company in the UK market and the largest nuclear power plant operator in the world, with 58 stations in France. EDF has not invested in a new nuclear station in more than a decade, but recently announced a new project in its home market. It is planning to build a new PWR at a site in Flamanville where two PWRs are already in operation.

Two of the UK's other large power companies also have nuclear experience. E.ON, for example, wholly or partly owns seven nuclear plants in Germany, while RWE, owner of npower and other UK companies, has five German stations. Vattenfall, the very large Scandinavian utility, has nuclear interests in Germany and in Sweden and in 2005 considered taking a stake in the UK market by buying Scottish Power.

These pan-European utilities have the experience and the capital strength that would be required to invest in new nuclear, in Europe as well as in the UK.

7.11 New designs on offer to the UK

The UK currently has a stake in two reactor designs but is in the process of disposing of its interest. One is the Pebble-Bed reactor, whose design and development are now being spearheaded by Eskom in South Africa, which already has one nuclear power plant at Keoberg. The Pebble-Bed reactor is very closely based on high-temperature gas reactor designs, with small spherical fuel elements made of graphite and uranium. The Pebble Bed is at the detailed design phase and as such is unlikely to be on the drawing board for a new reactor in the UK.

The UK's second stake in a new reactor is via BNFL's Westinghouse subsidiary, in the form of the AP600 and its larger variant the AP1000. However, the government announced in 2005 that it would sell off Westinghouse to Toshiba of Japan, and with it the UK's stake in the AP1000. The design has already been licensed by the US government – one beneficiary of a US programme to complete and license new reactor designs so they can be built quickly when power utilities decide it is time to invest. However, no AP1000 has yet been built and nuclear utilities, like any industry, prefer not to pioneer new designs. It has not yet been licensed by a European government, but as a variant of a well-known existing reactor type it is a potential contender to be built in the UK.

Atomic Energy of Canada is offering a new CANDU design rated at 700 MWe. This reactor has still to be built, although the CANDU technology is being aggressively promoted in the UK AECL, especially its new ACR-1000 model. Detailed design work on that reactor is still under way.

The two other candidates for the UK's next generation of nuclear reactors are the most likely contenders, which have the immense advantage that by the time the UK is ready to place an order for a new plant both will be in operation. Farthest ahead is General Electric's Advanced BWR (ABWR). This reactor is already in service in Japan and more are planned in that country. The UK has no experience with boiling-water reactors, except for trial reactor designs, which use similar principles (see Chapter 2). E.ON, RWE and Vattenfall, however, all have experience of BWRs.

The design that may be in the lead is the European Pressurised-Water Reactor (EPR), which is now being built at Olkiluoto in Finland. The benefit of the EPR is that it was designed from the start to satisfy the regulatory standards of more than one country – initially the French and German regulators but more recently their Finnish counterparts, as the first example of this design is being built there. That is not to say it would meet UK requirements unchanged, but as European regulators within the EU have harmonised their standards – if not their processes – the chance of major changes to the design being required may be much reduced.

The EPR is also planned for construction as France (and EDF's) next nuclear plant at Flamanville, so if one was ordered for the UK it should be able to rely on having at least two power stations ahead of it to provide operating experience.

All these reactors are described in more detail in the following chapter.

Chapter 8

New reactor designs

8.1 The Sizewell design

The UK's most recent reactor construction project was Sizewell B, a twin-unit pressurised-water reactor (PWR) based on Westinghouse's standard nuclear unit power plant system (Snupps). Sizewell B was ordered in 1987 and construction was completed in 1995, but when the reactor was ordered several other reactors based on the Snupps design were already in operation in the USA. The design dates back more than 20 years and is very unlikely to be employed again, especially as new designs offer better efficiency (and hence economics) and generally produce much less spent fuel and other radioactive waste over their lifetime. No other designs have been licensed in the UK and so far the licensing agency, the Nuclear Safety Directorate (part of the Health and Safety Executive) says that it has not been asked to begin formal examination of any designs.

If the decision was taken to go ahead with a new reactor or series of reactors in the UK, there are several designs that are clear front runners. The industry is at pains to point out that these are 'evolutionary' designs, based on familiar designs that have been in operation for up to 30 years in several countries. But while they are based on the same fundamental principles as existing pressurised-water reactors, boiling-water reactors, CANDUs and high-temperature reactors, the designs have been re-examined from the ground up. The aim is to increase safety and efficiency by simplifying the designs, while retaining all the necessary safety systems and ensuring that there is sufficient 'redundancy' to ensure safety. The principle of redundancy requires that there are backup or alternative systems for all the plant's safety functions. Such systems must also be 'diverse' – separated in space and using different technologies to fulfil their functions – to ensure they cannot suffer a 'common mode' failure. For example, in early designs it was common to have diverse shutdown systems, but the connections for those two systems might be allowed to pass through the same cable routes or even use the same cable trays. This meant a fire in that area could put both systems out of action and it later resulted in lots of on-site alterations or changes in design. New reactors have the principles of diversity and separation as a design basis, and realising it is made much easier by computer aided design techniques.

There are four leading contenders for new reactor designs in the UK. These 'most eligible' designs for the UK (see Chapter 7) have in some cases been built overseas, and in others been licensed in one country or more (i.e. passed a detailed inspection of the reactor design by the national safety authority). Some reactor designs have also been 'pre-certified' by the US regulator, the Nuclear Regulatory Commission. In the past the US NRC did not certify reactors until a site was chosen, as the site's characteristics have some part to play in the final reactor design (e.g. in the provision of cooling water, seismic qualifications, etc.). Pre-certification is intended to reduce site development times by resolving major issues in the basic design in advance. None of the overseas licensing or construction work guarantees that a reactor will successfully complete the UK's licensing process, although issues that might have been stumbling blocks in a UK licensing process may already have been resolved elsewhere.

The four designs are:

EPR. The European Pressurised-Water Reactor (EPR) is being developed by France and Germany. The EPR is a conventional, though advanced, PWR in which components have been simplified and considerable emphasis is placed on reactor safety. The design is now being built in Finland with a target completion during 2010. The French government also proposes building an additional EPR at Flamanville 3 in France. Present French policy suggests that additional EPRs might replace additional commercial reactors now operating in France starting in the late 2010s. The EPR was bid in early 2005 in competition to the AP1000 for four reactors at two sites in China. The proposed size for the EPR has varied considerably over time but might be around 1,600 MWe. Earlier designs were as large as 1,750 MWe. In either case the EPR would be the largest design now under consideration. The US arm of Areva, Framatome ANP, announced that the reactor was undergoing pre-certification in the USA in 2005. US utility Duke Power is evaluating the EPR, along with the AP1000 and ESBWR, for a combined operating and design licence in a process that began during 2005. A formal licence application by Duke will not occur for several years, though design selection might occur earlier.

AP600. The AP600 is a 600 MW PWR developed by Westinghouse BNFL that has already been certified by the NRC. The AP600, while based on previous PWR designs, has innovative passive safety features that permit a greatly simplified reactor design. Simplification has reduced plant components and should reduce construction costs. The AP600 has been bid overseas but has never been built. Westinghouse has de-emphasised the AP600 in favour of the larger, though potentially less expensive (on a per kilowatt basis) AP1000 design. The most recent AP1000 design has been bid in China with a 1,175 MW capacity. The AP1000 is an enlargement of the AP600, designed to almost double the reactor's target output without proportionately increasing the total cost of building the reactor. Westinghouse anticipates that operating costs will be below the average of reactors now operating in the United States. While Westinghouse BNFL owns rights to several other designs, the AP1000 is the principal

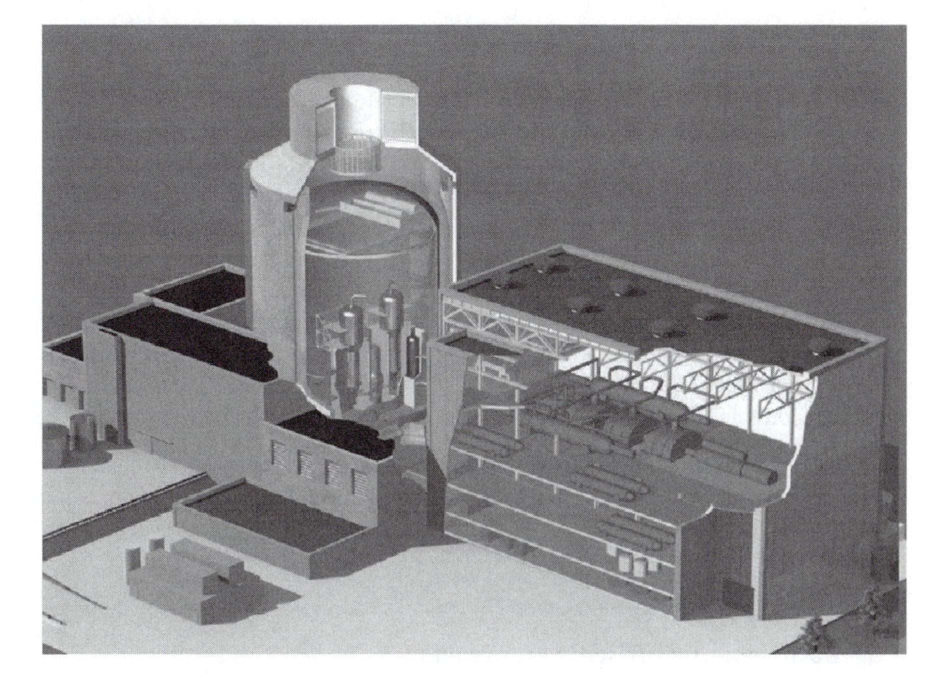

Figure 8.1 Cutaway of the AP1000 [Westinghouse]

product that the company now promotes in the United States for near-term construction. The AP1000 is a PWR with innovative, passive safety features and a much simplified design intended to reduce the reactor's material and construction costs while improving operational safety. One consortium of nine utilities called NuStart Energy promotes the AP1000 in the United States and has informed the NRC that it intends to apply for a combined construction and operating licence (COL) for the design. This is not a commitment to build the design.

Advanced BWR. The ABWR design has been both licensed in the USA and built elsewhere. Three ABWRs operate in Japan, and three are under construction, two in Taiwan and one in Japan. While the ABWR design is usually associated in the USA with General Electric, the Japanese units now being built are from Toshiba and Hitachi, which frequently associate with General Electric in possible ABWR projects in the US. There are many variations in ABWR design. The most frequently mentioned capacities are in the 1,250–1,500 MWe range though smaller and larger designs have been proposed depending on the vendor. Vendors now claim costs for building the ABWR that are low enough that they have attracted some customer interest.

ACR-700. Atomic Energy of Canada's 'Advanced CANDU Reactor' is considered by its vendor to be an evolution from AECL's internationally successful CANDU line of PHWRs. CANDU reactors and their Indian derivatives have been more of a commercial success than any other line of power reactors except the LWRs. One of the innovations in the ACR-700, compared to earlier CANDU designs, is that heavy water is used only as a moderator in the reactor. Light water is used as the coolant. Earlier CANDU designs used heavy water both as a moderator and as a coolant. This change makes it debatable whether the ACR-700 is a PHWR, a PWR or a hybrid between the two designs. AECL has aggressively marketed the ACR-700 offering low prices, short construction periods and favourable financial terms. AECL has subsequently slowed its efforts to certify the ACR-700 in the United States, although the firm still intends to follow the certification process at some point. AECL announcements indicate increased interest in a larger ACR-1000 design. If the scale economies attributed by Westinghouse BNFL to its AP series and by GE to its ABWR/ESBWR series are valid, one might anticipate parallel, cost-lowering results for the ACR series. Promised construction times for the ACR-700, as short as three years, would set modern records for large-reactor completion.

The following information about these four reactor types is based on vendor information. Neither the AP1000 nor the ACR-700 has yet been built, and although work has begun on the first EPR in Finland, at the time of writing the project is at an early stage.

8.2 European pressurised-water reactor

The EPR was designed to meet the requirements of European utilities set out in a 'European Utility Requirements' document, as well as a 'Utility Requirements Document' issued by the US's Electric Power Research Institute. It also complies with the recommendations (1993) and positions on major issues (1995) that were set up jointly by French and German safety authorities.

The technical guidelines covering the EPR design were validated in October 2000 by France's standing group of experts in charge of reactor safety (the advisory committee for reactor safety to the French safety authority) supported by German experts.

The EPR is the direct descendant of the N4 and Konvoi reactors, developed by Framatome and Siemens KWU respectively, which are themselves derivatives of standard US-type PWRs.

Thanks to an early focus on economic competitiveness during its design process, the vendor believes that the EPR has significantly reduced power generation costs, estimated to be 10 per cent lower than those of the most modern nuclear units currently in operation.

This high level of competitiveness would be achieved through:

- Unit power in the 1,600 MWe range (the highest unit power to date),
- A 36–37 per cent overall thermal efficiency, depending on site conditions (presently the highest thermal efficiency for any light-water reactors),
- A shortened construction time,
- A 60-year design life,
- Enhanced and more flexible fuel utilisation,

Figure 8.2 *Cutaway of the European pressurised-water reactor (EPR) design [published with permission of AREVA NP]*

- An availability factor up to 92 per cent, on average during the entire service life of the plant, obtained through long irradiation cycles, shorter refuelling outages and in-operation maintenance.

8.2.1 Construction schedule

Provisions have been made in the design, construction, erection and commissioning methods to shorten the EPR construction schedule. For example, the general layout of the main safety systems in four trains housed in four separate buildings simplifies, facilitates and shortens performance of the erection tasks for all work disciplines.

Location of electromechanical equipment at low levels means that it can be erected very early on in the programme, thus shortening the critical path of the construction schedule.

Three main principles are applied to the EPR construction and erection: minimisation of the interfaces between civil works and erection of mechanical components; modularisation; and piping pre-fabrication. Optimising the interfaces between civil and erection works results in the implementation of a construction methodology 'per level' or with 'grouped levels'. That means that it is possible to carry out equipment and system installation work at level 'N' where civil construction is finished, while at the same time finishing construction works at the level above ('$N + 1$') and carrying out major construction work at the two levels above ('$N + 2$' and '$N + 3$')

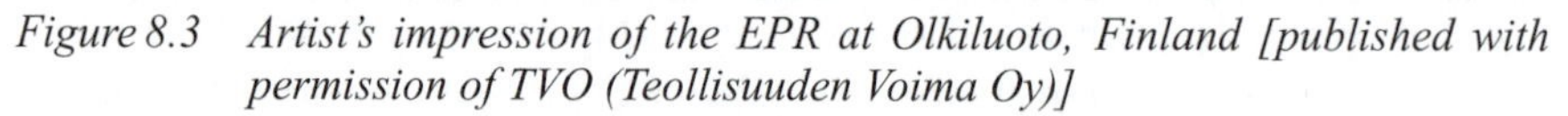

Figure 8.3 Artist's impression of the EPR at Olkiluoto, Finland [published with permission of TVO (Teollisuuden Voima Oy)]

simultaneously; this methodology is used for all the different buildings except for the reactor building.

8.2.2 Operation, maintenance and services

The key to ensuring an economic return from the nuclear reactor is keeping it in operation: that means that scheduled outages (for refuelling and maintenance) should be minimised and unscheduled outages should be reduced, if possible, to zero.

The EPR and its equipment and systems have been designed to allow for efficient refuelling outages and to simplify and optimise inspection and maintenance in order to increase plant availability and reduce maintenance costs, two major objectives of plant operators.

The EPR is designed to reach 92 per cent availability over the entire 60 years of its design lifetime. This is made possible by short scheduled outages for fuel loading or unloading and for in-service inspections and maintenance, and also through reduced downtimes attributable to unscheduled outages.

Moreover, the reactor building is designed to be accessible, under standard safety and radiation protection conditions, while the reactor is at power. This development enables outage and maintenance operations to be prepared before the scheduled outage begins, reducing the length of the outage. Access to the reactor building during power operation allows preventive maintenance and refuelling tasks to begin up to seven days before reactor shutdown and can be continued for up to three days after reactor restart.

The possibility of gaining access throughout the reactor building with the reactor online also makes it easier to carry out servicing and some maintenance if they are required outside scheduled outage periods.

The duration of the plant shutdown phase is reduced by decreasing the time required for reactor coolant system cooldown, depressurisation and vessel head opening. Similarly the length of the restart is reduced: the time needed to run the beginning-of-cycle core physics tests has been reduced. Durations of about 70 and 90 h are scheduled for the shutdown and restart phases, respectively. For the fuel loading/unloading operations, a period of about 80 h is scheduled.

Thanks to these changes, the duration of a regular outage for preventive maintenance and refuelling should be reduced to 16 days, while a refuelling outage can be reduced to 12 days. Ten-year outages for main equipment in-service inspection, turbine overhaul and containment pressure test are planned to last 38 days.

8.2.3 Operational flexibility

In terms of operation, the EPR is designed to offer the utilities a high level of flexibility. It can be operated over the long term at any power level between 20 per cent and 100 per cent of its nominal power, with the primary and secondary frequency controls in operation.

This allows it to respond to scheduled and unscheduled power grid demands for load variations, help manage grid perturbations and mitigate grid failures.

The EPR's core design and higher overall efficiency, compared to the reactors in operation today, also claims the following advantages:

- 17 per cent saving on uranium consumption per produced megawatt hour,
- 15 per cent reduction on long-lived actinides generation per megawatt hour,
- 14 per cent gain on the 'electricity generation' versus 'thermal release' ratio (compared to 1,000 MWe-class reactors),
- The ability to use mixed oxide fuel.

8.2.4 Safety systems

The safety design of the EPR builds on the experience of other French and German reactors.

The design is simplified by separating operating and safety functions. Four-fold redundancy is applied to the safeguard systems and to their support systems. This architecture allows one system to undergo maintenance during plant operation without requiring the plant to be shut down, as there is still three-fold redundancy, thus helping to ensure a high plant availability factor.

The four trains of the safety systems are located in four different buildings in which strict physical separation is applied. Each has systematic functional diversity.

The emergency power supply system is designed to meet the requirements of the $N+2$ concept (i.e. assuming a single failure on one train and a maintenance operation on another).

8.2.4.1 Buildings and components

The reactor building houses the main equipment of the nuclear steam supply system (NSSS) and the in-containment refuelling water storage tank (IRWST). Its main function is to ensure protection of the environment against internal and external hazard consequences under all circumstances. It consists of a cylindrical pre-stressed inner containment with a metallic liner surrounded by an outer reinforced concrete shell. The main steam and feedwater valves are housed in dedicated reinforced concrete compartments adjacent to the reactor building.

In the primary system:

- The pressuriser is located in a separate area,
- There are concrete walls between the loops and between the hot and cold legs of each loop,
- There is a concrete wall (the secondary shield wall) around the primary system to protect the containment from missiles and to reduce the spread of radiation from the primary system to the surrounding areas.

The fuel building, located on the same common basemat as the reactor building and the safeguard buildings, houses the fresh fuel, the spent fuel in an interim fuel storage pool and associated handling equipment. Operating compartments and passageways, equipment compartments, valve compartments and the connecting pipe ducts are separated within the building. Areas of high activity are separated from areas of low activity by means of shielding. The mechanical floor houses the fuel pool cooling system, the emergency boration system and the chemical and volume control system. The redundant trains of these systems are physically separated by a wall into two building parts.

Two diesel buildings house the four emergency diesel generators and their support systems and supply electricity to the safeguard trains in the event of a complete loss of electrical power. The physical separation of these two buildings provides additional protection.

To withstand a major earthquake, the entire nuclear island stands on a single thick reinforced concrete basemat. The building height has been minimised and heavy components and water tanks are located at the lowest possible level.

To withstand a large airplane crash, the reactor building, spent fuel building and two of the four safeguard buildings are protected by an outer shell made of reinforced concrete. The other two safeguard buildings are protected by geographical separation. Similarly, the diesel generators are located in two geographically separate buildings to avoid common failures.

8.2.5 Primary system

The EPR primary circuit is a four-loop design (i.e. it has four coolant loops each with a steam generator and associated pumps).

The reactor pressure vessel, pressuriser and steam generators feature larger volumes than similar components from previous designs to provide additional benefit in terms of operation and safety margins.

The increased free volume in the reactor pressure vessel, between the nozzles of the reactor coolant lines and the top of the core, provides a higher water volume above the core and thus additional margin with regard to the core 'dewatering' time in the event of a postulated loss of coolant accident.

This increased volume would also be beneficial in shutdown conditions in case of loss of the residual heat removal system function.

Larger water and steam phase volumes in the pressuriser smooth the response of the plant to normal and abnormal operating transients, allowing the operator more time to counteract accident situations and extend equipment lifetime.

The larger volume of the steam generator secondary side results in increasing the secondary water inventory and the steam volume, which offers several advantages. Due to the increased mass of secondary side water, in case of an assumed total loss of the steam generator feedwater supply, the dry-out time would be at least 30 min, sufficient time to recover a feedwater supply or to decide on other countermeasures. In addition, the primary system design pressure has been increased in order to reduce the actuation frequency of the safety valves, which is also an enhancement in terms of safety.

8.2.5.1 Pressure vessel

To minimise the number of large welds, and consequently reduce their manufacturing cost and time for in-service inspection, the upper part of the RPV is machined from one single forging and the flange is integral to the nozzle shell course. The reduced number of welds and the weld geometry decrease the need for in-service inspection, facilitate non-destructive examinations and reduce inspection duration as well. Nozzles of the set-on type facilitate the welding of the primary piping to the RPV and the welds' in-service inspection as well.

The lower part of the RPV consists of a cylindrical part at the core level, a transition ring and a spherical bottom piece. As the in-core instrumentation is introduced through the closure head at the top of the RPV, there is no penetration through the bottom of the vessel. The elimination of any penetration through the RPV bottom head strengthens its resistance in case of postulated core meltdown and prevents the need for in-service inspection and potential repairs.

The RPV has been designed for non-destructive testing during in-service inspections. In particular, its internal surface is accessible to allow 100 per cent visual and/or ultrasonic inspection of the welded joints from the inside.

Consistently with the EPR's 60-year design life, there is an increased margin with regard to reactor pressure vessel embrittlement. The ductile–brittle transition temperature of the vessel material remains lower than 30 °C at the end of the design life, achieved by the choice of material and its specified low content in residual impurities, and also thanks to a reduced neutron fluence due to the inclusion of a neutron reflector surrounding the core.

8.2.5.2 Steam generators

The steam generator is an enhanced version of the axial economiser steam generator implemented on N4 plants. The axial economiser allows for higher steam pressure output compared to a conventional design, without impairing access to the tube bundle for inspection and maintenance. The very high steam saturation pressure at tube bundle outlet (78 bar) is a major contributor to the high efficiency of the EPR (37 per cent).

The increase in steam volume and the set pressure of the secondary safety valves prevent any liquid release to the environment in case of steam generator tube rupture. Particular attention was given during the design of the EPR steam generator to cancel out secondary cross-flows to protect the tube bundle against vibration risks.

The steam drum volume has been augmented. This feature, plus a safety injection pressure lower than the set pressure of the secondary safety valves, would prevent the steam generators from filling up with water in case of steam generator tube rupture to avoid liquid releases.

8.2.5.3 Reactor coolant pumps

The reactor coolant pumps provide forced circulation of water through the reactor coolant system. This circulation removes heat from the reactor core to the steam generators, where it is transferred to the secondary system.

A reactor coolant pump is located between the steam generator outlet and the reactor vessel inlet of each of the four primary loops. The pump design is an enhanced version of the model used in the N4 reactors. The pump capacity has been increased to comply with the EPR operating point. In addition, a new safety device, a standstill seal, has been added as shaft seal back-up. It ensures that even in case of total station blackout or failure of the main seals no loss of coolant would occur.

8.2.5.4 Pressuriser

The pressuriser has a larger volume to smooth operating transients, in order to:

- ensure the equipment's 60-year design life,
- increase the time available to counteract an abnormal operating situation,
- make maintenance and repair easier and reduce radiological doses.

A dedicated set of valves for depressurising the primary circuit is installed on the pressuriser, in addition to the usual relief and safety valves, to prevent the risk of high-pressure core melt accident.

8.2.5.5 Safety injection system

The safety injection system (SIS/RHRS) performs a dual function both during the normal operating conditions in residual heat removal mode and in the event of an accident. The system consists of four separate and independent trains, each providing the capability for injection into the coolant system.

During normal operating conditions, the system:

- transfers heat from the reactor to the Component Cooling Water System (CCWS) when heat transfer via the steam generators is no longer sufficiently effective,

- transfers heat continuously from the coolant system or the reactor refuelling pool to the CCWS during cold shutdown and refuelling shutdown, as long as any fuel assemblies remain inside the containment.

In the event of an assumed accident and in conjunction with the CCWS and the Essential Service Water System (ESWS), the SIS in RHR mode maintains the RCS core outlet and hot leg temperatures below 180 °C following a reactor shutdown.

The four redundant and independent SIS/RHRS trains are arranged in separate divisions in the safeguard buildings. Each train is connected to one dedicated coolant loop and is designed to provide the necessary injection capability required to mitigate accident conditions. This configuration greatly simplifies the system design.

The design also makes it possible to have extended periods available for carrying out preventive maintenance or repairs. For example, preventive maintenance can be carried out on one complete safety train during power operation.

Within the containment the in-containment refuelling water storage tank (IRWST) contains a large amount of borated water and collects water discharged inside the containment. Its main function is to supply water to various systems, and to flood the spreading area in the event of a severe accident.

The EPR complies with the safety objectives set up jointly by the French and German safety authorities for future PWR power plants. They include further reduction of core melt probability, practical elimination of accident situations which could lead to large early release of radioactive materials and the requirement for very limited protective measures in the case of a core melt – that is, no permanent relocation, no need for emergency evacuation outside the immediate vicinity of the plant, limited sheltering and no long-term restriction in the consumption of food.

The protection and safeguard actions needed in the short term in the eventuality of an incident or accident are automated. Operator action is not required before 30 min for an action taken in the control room, or 1 h for an action performed locally on the plant.

The increase in the volumes of the major components (reactor pressure vessel, steam generators, pressuriser) gives the reactor extra inertia which helps to extend the time available to the operators to initiate the first actions. The choices taken for the installation of the safeguard systems and the civil works minimise the risks arising from the various hazards (earthquake, flooding, fire, aircraft crash).

The safeguard systems are designed on the basis of a quadruple redundancy, both for the mechanical and electrical portions and for the I&C. This means that each system is made up of four sub-systems, or 'trains', each one capable by itself of fulfilling the whole of the safeguard function. The four redundant trains are physically separated from each other and geographically shared among four independent divisions (buildings). Each division includes:

- a low-head injection system and its cooling loop, together with a medium-head injection system, for borated-water safety injection into the reactor vessel in the case of a loss of coolant accident,

- a steam generator emergency feedwater system,
- the electrical systems and I&C linked to these systems.

The building housing the reactor, the building in which the spent fuel is interim-stored and the four buildings corresponding to the four divisions of the safeguard systems are given special protection against externally generated hazards such as earthquakes and explosions.

This protection is further strengthened against an airplane crash. The reactor building is covered with a double concrete shell: an outer shell made of 1.30 m thick reinforced concrete and an inner shell made of pre-stressed concrete and also 1.30 m thick which is internally covered with a 6 mm thick metallic liner. The thickness and the reinforcement of the outer shell on its own have sufficient strength to absorb the impact of a military or large commercial aircraft. The double concrete wall protection is extended to the fuel building, two of the four buildings dedicated to the safeguard systems, the main control room and the remote shutdown station which would be used in a state of emergency.

8.2.5.6 Limiting severe accident consequences

In response to the new safety model for the future nuclear power plants, introduced as early as 1993 by the French and German safety authorities, the plant design must be such that a core melt accident, although highly unlikely, causes only very limited off-site measures in time and space.

In addition to the usual reactor coolant system depressurisation systems on the other reactors, the EPR is equipped with valves dedicated to preventing high-pressure core melt in the eventuality of a severe accident. These valves would then ensure fast depressurisation, even in the event of failure of the pressuriser relief lines.

Their relieving capacity guarantees fast primary depressurisation down to values of a few bars, precluding any risk of containment pressurisation through dispersion of corium debris in the event of vessel rupture.

The high mechanical strength of the reactor vessel is sufficient to rule out its damage by any reaction, even high-energy, which could occur on the inside between corium and coolant.

The portions of the containment with which the corium would come in contact in the eventuality of a core melt exacerbated by ex-vessel progression – namely the reactor pit and the core spreading area – are kept 'dry' (free of water) in normal operation. Only when it is spread inside the dedicated area, therefore already partially cooled, surface-solidified and less reactive, would the corium be brought into contact with the limited water flow intended to cool it down further.

In the case of a severe accident, hydrogen would be released in large quantities inside the containment. This would happen first of all by reaction between the coolant and the zirconium which is part of the composition of the fuel assembly claddings, then, in the event of core melt and ex-vessel progression, by reaction between the corium and the concrete of the corium spreading and cooling area.

For this reason, the pre-stressed concrete inner shell of the containment is designed to withstand the pressure which could result from the combustion of this hydrogen. Further, devices called catalytic hydrogen recombiners are installed inside the containment to keep the average concentration below 10 per cent at all times, to avoid any risk of detonation.

8.2.5.7 Collection of inter-containment leaks

In the eventuality of a core melt leading to vessel failure, the containment remains the last of the three containment barriers; this means that provisions must be taken to make sure that it remains undamaged and leak-tight. For the EPR, the following measures have been adopted:

- a 6 mm thick metal liner internally covers the pre-stressed concrete inner shell,
- the internal containment penetrations are equipped with redundant isolation valves and leak recovery devices to avoid any containment bypass,
- the architecture of the peripheral buildings and the sealing systems of the penetrations rule out any risk of direct leakage from the inner containment to the environment,
- the space between the inner and outer shells of the containment is passively kept at slight negative pressure to enable the leaks to collect there,
- these provisions are supplemented by a containment ventilation system and a filter system upstream of the stack.

8.3 Westinghouse AP600/AP1000

In the 1980s, Westinghouse embarked on a development programme for its PWR technology. Its objectives were to improve availability, generation economics, and operation and maintenance.

The result was the AP600, a 600 MWe PWR that is intended to use field-proven technology from the existing PWR fleet, but also to use 'passive' safety systems and extensive plant simplifications. The design integrates modifications based on lessons learned or operating experience, and incorporates changes only where clear benefit is measurable.

The AP600 NSSS is based on the standard two-loop Westinghouse PWR designs that have collectively logged more than 100 reactor years of operation. US plants of this design include Ginna, Prairie Island, Kewaunee and Point Beach. Korea's Kori 2 plant, which became operational in 1983, is a Westinghouse two-loop PWR that has been among the best performers in the world, and it is this plant design which is the basis, or reference, for the AP600 NSSS design.

8.3.1 Passive safety

The AP600 passive safety systems do not rely on active management by the operating staff or external power. They include passive safety injection, passive residual heat

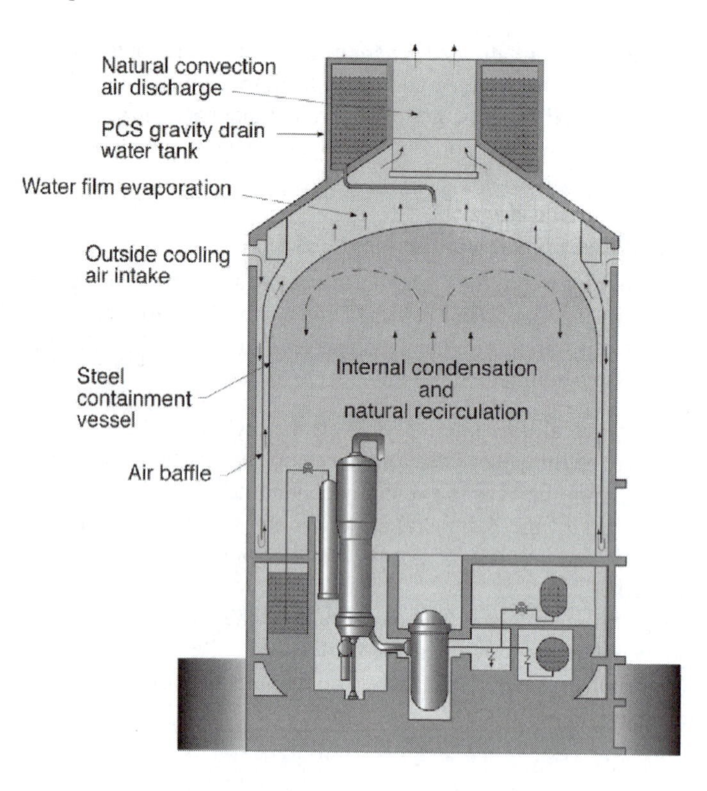

Figure 8.4 Westinghouse's AP600 uses 'passive' cooling [Westinghouse]

removal, passive containment cooling, and passive main control room habitability maintenance. These systems employ natural forces and stored energy to operate. They are highly reliable because in the event of an accident, with an assumed unavailability of non-safety systems, they do not require motors, pumps or diesel generators to be started.

Several aspects of the passive safety systems have been used in existing nuclear plants. The accumulators are a part of most PWR designs, so their use is well understood. Several early boiling-water reactors (BWRs), like Dresden in the US, used isolation condensers as natural circulation closed-loop heat removal systems. The AP600 passive residual heat removal heat exchanger was designed with the benefit of this experience.

BWRs have used automatic depressurisation systems (ADS) and spargers (sprinkler systems) for many years. Use of slow-opening valves is a result of understanding the air clearing loads discovered in BWR operation. The AP600 ADS incorporates spargers to allow depressurisation of the reactor coolant system to the in-containment refuelling water storage tank (IRWST) in lieu of the containment atmosphere to minimise the containment clean-up following an ADS actuation. The sparger design incorporates BWR design and operating experience.

8.3.2 Reactor design

The core, reactor vessel and internals of the AP600 are essentially those of a conventional Westinghouse PWR design. Several important features, all based on existing technology, have been used to improve the performance characteristics of the AP600. They include ring forgings to eliminate vertical weld seams, location of circumferential welds outside high neutron flux beltline region and control of the chemical composition of the vessel material to reduce irradiation damage.

Although the AP600 reactor has two primary coolant loops, the reactor vessel is similar to the standard three-loop model. Each primary loop has two cold leg nozzles and one hot leg nozzle, so the standard three-loop vessel accommodates the AP600 nicely with slight re-orientation of the nozzles. Moreover, the larger reactor vessel on the AP600 accommodates a larger reactor core than that for a traditional two-loop plant. With the larger reactor core and the same power rating, the core power density is reduced.

A radial neutron reflector surrounds the core. This reflector reduces neutron leakage, thereby improving core neutron utilisation and reducing the damaging neutron fluence on the reactor vessel. This contributes to lowering fuel cycle cost and extending reactor life. The reduced fluence is important in view of the 60-year design objective of the AP600.

The combination of the radial reflector, the low power density core and optimised fuel assemblies results in a 20 per cent fuel cycle cost saving, compared to a standard PWR design of the same power rating. The core design allows 18-month refuelling cycles to be achieved with an 85 per cent capacity factor (approximately 466 effective full power days per cycle) and requires no burnable absorbers other than for the first cycle of operation.

8.3.3 Fuel and core

The AP600 fuel design is based on standard 17 × 17 optimised fuel technology currently being used in approximately 120 operating plants worldwide. Over 25,000 17 × 17 fuel assemblies have already been manufactured. Fuel performance improvements, such as zircaloy grids, removable top nozzles and longer burnup, have been added.

The advantages of the AP600's low power density are achieved by making the core larger than conventional 600 MWe designs. As a result, the number of fuel assemblies is increased from 121 to 145, so that many of the important nuclear and thermal parameters are improved by 25 to 30 per cent over those of a standard plant of the same power rating. This results in lower fuel enrichments, less reliance on burnable absorbers, longer achievable fuel cycles and an increase of more than 15 per cent in some safety margins. A larger gas plenum has been incorporated into the fuel to allow higher burnup.

8.3.4 Load following

Another core design feature of the AP600 is the use of reduced worth control rods, which are termed 'grey' rods, to achieve daily load follow without requiring daily changes in the soluble boron concentration. While the same control rods are used in existing PWRs for load follow, it is also necessary for these plants to process thousands of gallons of water per day in order to change the soluble boron concentration sufficiently to achieve a daily load follow schedule. The use of grey rods eliminates the need for processing the primary coolant on a daily basis and greatly simplifies operations using the boron systems.

With the exception of the neutron absorber materials used, the design of the grey rod assembly is identical to that of a normal control rod assembly. Thus, the design of grey rods, the fuel assembly, the reactor vessel and reactor internals are all based on existing Westinghouse PWR technology.

8.3.5 Reactor coolant system

The AP600 reactor coolant system (RCS) consists of two heat transfer circuits, each with a steam generator; two reactor coolant pumps; and a single hot leg and two cold legs for circulating reactor coolant between the reactor and the steam generators. The system also includes a pressuriser, interconnecting piping, and the valves and instrumentation necessary for operational control and safeguards actuation. All equipment is located in the reactor containment and uses major components that have been proven in operating reactors under similar flow, temperature and pressure conditions.

The reactor coolant system employs high inertia, high reliability, low maintenance canned motor pumps for circulating primary reactor coolant through the reactor core, piping and steam generators.

Two pumps are mounted directly in the channel head of each steam generator. This allows the pumps and steam generator to use the same structural support, greatly simplifying the support system and providing more space for pump and steam generator maintenance. The combined steam generator/pump vertical support is a single column extending from the floor to the bottom of the channel head.

Inverted canned motors have been in operation for over 25 years in fossil boiler circulation systems with better operating reliability than upright units, because the motor cavity is self-venting into the pump casing, avoiding the potential for gas pockets in the bearing and water regions. Approximately 1,300 units have been built and placed into service.

The pumps are integrated into the steam generator channel head in the inverted position. The advantages of this pump design are significant. The auxiliary fluid systems needed to support a canned motor pump are much less complex than those needed for a shaft seal-type pump. The canned motor pump is more tolerant of off-design conditions than the shaft seal pump and inherently reduces the potential for small loss of coolant accidents by eliminating the shaft seal. The integration of the pump suction into the bottom of the steam generator channel head eliminates the crossover leg of coolant loop piping, reduces the loop pressure drop, simplifies the foundation and support system for the steam generator, pumps and piping, and eliminates the potential for the core to become uncovered during a small loss of coolant accident.

8.3.6 Instrumentation and control

The AP600 instrumentation and control (I&C) system is based on existing digital technology and hardware, which ensures greater design flexibility, improves plant safety, availability, reliability and maintainability, and dramatically reduces the quantity of plant cabling. Included in the I&C systems are the main and emergency control boards, the plant protection system, the NSSS control systems, the turbine-generator control system, the balance of plant control systems and the plant-wide monitoring systems.

8.3.7 Steam generators

Two model Delta-75 steam generators are used in the AP600. The Delta-75 steam generator is based on standard Westinghouse Model-F technology. The Model-F is a proven design with approximately 75 units currently in commercial operation with less than one tube plugged per steam generator for every four years of operation.

These reliability achievements are due to a variety of design enhancements incorporated to improve performance and increase service life. Enhancements include full-depth hydraulic expansion, stainless steel broached tube support plates, the use of thermally treated Inconel 690 (I-690) for the tube material to improve corrosion resistance, the addition of upgraded anti-vibration bars to reduce wear, upgraded primary and secondary moisture separators, and use of a triangular tube pitch.

I-690 has excellent overall corrosion resistance, is resistant to primary water-stress corrosion cracking, and has low primary release rates which are expected to reduce primary side radiation levels by at least 50 per cent compared to units employing I-600 tube material.

The Delta-75 steam generator used for the AP600 is currently in use at the V.C. Summer plant in the US and has completed one 18-month cycle of operation with the new steam generators. The design enhancements incorporated into the Delta-75 steam generators have provided the plant with greater operating flexibility and increased safety margins.

8.3.8 Pressuriser

The AP600 pressuriser is based on the standard Westinghouse design currently used in approximately 70 operating plants worldwide. Westinghouse pressurisers are a technology with over 30 years of successful operating experience. The AP600 pressuriser is 1,600 cubic feet, which is about 30 per cent larger than would normally be used in a plant of comparable power rating.

The larger pressuriser increases transient operation margins and eliminates the need for relief valve actuation, which results in a more reliable plant that will have fewer reactor trips and that will require less time to recover after a transient. It also eliminates the need for power-operated relief valves, which eliminates a possible source of RCS leakage and maintenance.

The AP600 pressuriser heaters are similar in design to those employed in operating Westinghouse PWRs. The heaters are vertically mounted, extending up through penetrations in the bottom head of the pressuriser shell. They are also individually seal-welded to the penetrations providing the system pressure boundary. The pressuriser heaters are one of the many components that have achieved very good overall performance in operating plants.

8.4 Advanced BWR

The design for the Advanced BWR (ABWR) differs from most operating BWRs in Europe and the US in a number of ways. The ABWR's building volume is only about 70 per cent of more recent BWR buildings. This cuts construction time and cost, and it makes the design more rugged and more immune to earthquakes.

The ABWR reactor pressure vessel is 21 m high and 7.1 m in diameter. It is designed for a life of 60 years under normal operating conditions. Much of the vessel, including the four vessel rings from the core beltline to the bottom head, is made from single forging. The vessel has no nozzles greater than 2 inches in diameter anywhere below the top of the core because the external recirculation loops have been eliminated. Because of these two features, over 50 per cent of the welds and all of the pipings and pipe supports in the primary system have been eliminated, and along with them the biggest source of occupational exposure in the BWR.

The reactor vessel is designed to contain and support the core and fuel. A major design feature in the reactor vessel is the use of ring-forging. The ring-forging reduces the number of necessary rolled and welded plates used in previous vessel designs and overall complexity of the vessel. The vessel is fabricated from 508 steel forgings with controlled copper, nickel, sulphur and phosphorous in the beltline region of the vessel.

The reactor vessel is vertically mounted with an integral hemispherical lower head and a removable hemispherical upper closure head. The control element drive mechanism is welded upon the closure head. Some nozzles for coolant, borated water are provided on the reactor vessel. The reactor vessel assembly, including lower and upper head, nozzles and CEDM, is supported by four columns located under the vessel inlet nozzles. These columns are designed to restrain vertical motion during earthquakes and following branch line pipe breaks.

8.4.1 Maintenance and operation

Particular attention was paid to designing the plant for ease of maintenance. One of the features of the ABWR is that it has eliminated external recirculation systems. The external recirculation pumps and piping have been replaced by ten internal recirculation pumps mounted on the bottom head. These reactor internal pumps (RIPs) are improved versions of those used in Europe, for which there are over 1,000 pump years of operating experience.

Cobalt has been eliminated from the design. This is because in earlier designs activated cobalt became one of the major sources of radiological contamination – many existing plants have progressively replaced components containing cobalt. The steel used in the primary system is made of nuclear-grade material (low carbon alloys) resistant to intergranular stress corrosion cracking.

Additional control rod power supply improves reliability. In the current BWR, the emergency control rods are hydraulic. In the ABWR, they are electro-hydraulic. Having an additional drive mechanism reduces the probability of failure.

Fine motion control rod drives (FMCRD) are being introduced in the ABWR. Day-to-day operation is performed with an electric stepping motor, which moves the drive in very small increments. The control rods are scrammed hydraulically but can also be scrammed by the electric motor as a back-up. The FMCRD is so reliable that it is not necessary to inspect all of them during the lifetime of the plant. Therefore only three drives will be removed for inspection during each outage, which is a huge saving in time. The FMCRDs have continuous clean water purge to keep radiation at very low levels.

8.4.2 Instrumentation and control

The instrumentation and control (I&C) systems use digital and fibre optic technologies. The ABWR has four separate divisions of safety system logic and control, including four separate, redundant multiplexing networks to provide absolute assurance of plant safety. Each system includes microprocessors to process incoming sensor information and to generate outgoing control signals, local and remote multiplexing units for data transmission and a network of fibre optic cables. The controllers are fault tolerant, meaning that they continually generate signals to simulate input data and compare the result against the expected outcome. Controllers for both sensors and equipment are located on cards, which are remotely distributed. In the event that a problem is detected by the controller, a signal is sent to the control room. Within minutes, the malfunctioning card can be replaced with a spare.

All major equipment and components have been engineered with service and maintenance in mind, which will minimise downtime and reduce worker exposure to radiation.

Multiplexing and fibre optics have dramatically reduced the amount of cabling in the plant. This has another benefit: it shortens the critical path for the construction schedule by one month.

The entire plant can be controlled from one console. The panels in the centre control non-safety systems in the nuclear island, the ones on the left control the safety systems and the ones on the right the balance of plant. The CRTs allow the operator to call up any system, its sub-systems and components just by touching the screen. It is also possible to operate the entire system by means of a system master command.

The ABWR is designed to an envelope of site conditions that covers almost all of the available nuclear sites in the world, including the sites with high seismic potential.

The reactor and turbine buildings are arranged ‘in-line’ and none of the major facilities are shared with the other units. The containment is a reinforced concrete containment vessel (RCCV) with a leak-tight steel lining. The containment is surrounded by the reactor building, which doubles as a secondary containment. A negative pressure is maintained in the reactor building to direct any radioactive release from the containment into a gas treatment system. The reactor building and the containment are integrated to improve the seismic response of the building without an increase in the size and load-bearing capability of the walls.

Construction of the plant makes use of large modules that are pre-fabricated in the factory and assembled on site. A 1,000 ton crane lifts these modules and places them vertically into the plant. The use of modular and other construction techniques, plus the RCCV, reduces construction times from 66 to 50 months.

The ABWR has three completely independent and redundant divisions of safety systems. These systems are mechanically separated and have no cross connections as in earlier BWRs. They are electronically separated so that each division has access to redundant sources of ac power and, for added safety, its own dedicated emergency diesel generator. Divisions are physically separated. Each division is located in a different quadrant of the reactor building, separated by fire walls. A fire, flood or loss of power that disables one division has no effect on the capability of the other safety systems. Finally, each division contains both a high- and low-pressure system and each of these has its own dedicated heat exchanger to control core cooling and remove decay heat. One of the high-pressure systems, the Reactor Core Isolation Cooling (RCIC) system, is powered by reactor steam and provides the diverse protection needed should there be a station blackout.

The safety systems have the capability to keep the core covered at all times. Because of this capability and the thermal margins built into the fuel designs, the frequency of transients that will lead to a scram and therefore to plant shutdown has been greatly reduced. In the event of a loss of coolant accident, plant response has been fully automated and operator action is not required for 72 h, the same capability as for the passive plants.

ABWR features are intended to passively mitigate the consequences of a severe accident. One of these is a system which automatically floods the containment area below the reactor vessel should there ever be a core melt. The radiative heat of the core debris melts a fusible valve, thereby releasing water from the suppression pool in the ABWR. This quenches the core debris and limits the amount of non-condensable gas that is generated from the concrete–core reaction. The ABWR also includes a system to prevent catastrophic failure of the containment. When containment pressures near design limits, a rupture disk located in a hardened vent opens. This creates a path from the air space above the suppression pool to the atmosphere for steam and heat to be released until the operator manually closes valves in the vent. Most fission products are retained in the suppression pool.

The ABWR's operating cycle is 18 months with capability up to 24 months. The refuelling outages for these cycle lengths are 43 days. These outage lengths assume that there is only normal maintenance work and no major turbine generator work. A number of design features speed up the outage: an automated fuel movement platform, a 'minimum shuffle' core loading strategy, fewer control rod drive removals, automated handling of FMCRDs and RIPs and automated startup times, all of which improve the availability factor by about 2.5 per cent.

8.4.3 ABWRs in operation

Kashiwazaki-Kariwa 6, the first of two GE ABWRs in Japan, began generating electricity in January 1996. A second unit, Kashiwazaki-Kariwa 7, began commercial operation in mid-1997. Its design is similar to ABWRs certified in the United States; construction was completed in 52 months, 10 weeks ahead of schedule.

8.5 Advanced CANDU reactor

The Advanced CANDU Reactor (ACR) marks a shift in the CANDU design from using heavy-water coolant to using light water for cooling. Heavy water, used in the original CANDU design for both cooling and moderating the reaction, is now used only as a moderator. That means the ACR-700 design requires 75 per cent less heavy water than the previous version, CANDU 6.

Other similarities with PWR and BWR designs include the balance of plant systems (outside the NSSS) and the steam generators, but some differences remain.

The ACR retains the basic configuration of the CANDU design, with reactor pressure tubes in a horizontal configuration passing through a calandria. However, the distance between each fuel channel in the reactor core has been increased to improve access at the reactor face and make the design more amenable to reactor face installation, inspection and maintenance activities.

The ACR, like all previous CANDU designs, includes online fuelling, i.e. fuel is replaced at power and without shutting the plant down. Current light-water reactors shut down on an 18-month or 24-month cycle for refuelling, whereas the ACR design aims for a three-year operational cycle with a 21-day maintenance outage.

The ACR fuel is similar to that of other CANDU designs in that it is contained in short fuel bundles that are moved progressively through the reactor in the horizontal pressure tubes. The fuel is enriched to only around 2 per cent, compared to slightly higher enrichments in PWRs and BWRs, but this represents a change from the conventional CANDU design, which uses natural uranium fuel.

The other difference is that the ACR-700 does not have boron in the reactor coolant system, which ensures long-term plant life.

The ACR design has two independent safety shutdown systems; an inherent passive emergency fuel coolant capability, in which the moderator system can absorb excess heat; and limited challenge to containment barriers, even in the event of a severe accident. The ACR design incorporates further passive safety, such as a highly stable core design, large operating margins and long times available for operator action.

Due to data showing that 65–75 per cent of forced outage items are related to the balance of plant (the area outside the NSSS), AECL is working with Bechtel Hitachi to reduce that rate either by adding redundancies or by re-engineering equipment.

To reduce construction costs AECL is examining the plant design to standardise pipe sizes, valves, hangers, supports and other equipment across all systems of the plant. This will also help ease maintenance efforts and minimise spare part requirements.

AECL has decided to focus the ACR programme on the development of the ACR-1000 product as the lead design, moving on from the lower-power ACR-700. AECL's planned approach to new build in Ontario envisions the first ACR-1000 going into service in 2016, in which timeframe an energy gap is predicted to open up significantly. AECL says improvements in project engineering, manufacturing and construction technologies mean it can 'guarantee a significantly shorter schedule, while ensuring high quality design and construction, and reduced rework'.

8.6 Other designs

Several other reactor designs are in the certification or pre-certification phase in the USA or elsewhere, but are considered unlikely contenders for a new UK programme. They include:

8.6.1 Economic simplified, boiling-water reactor

The Economic Simplified, Boiling-Water Reactor (ESBWR) is a new, simplified BWR design promoted by General Electric and some allied firms. The ESBWR constitutes an evolution and merging of several earlier designs including the ABWR that are now less actively pursued by GE and other vendors beyond the exceptional case of Bellefonte in Alabama. The intent of the new design, which includes new passive safety features, is to cut construction and operating costs significantly from earlier ABWR designs. GE and others are investing heavily in the ESBWR though the design might not be available for deployment for several years. The nine-utility NuStart Energy group promotes the ESBWR as well as the AP1000 design. NuStart has informed the NRC that it intends to apply for pre-certification for the ESBWR in addition to any AP1000 application.

8.6.2 Pebble-bed modular reactor

The Pebble-Bed Modular Reactor (PBMR), which uses helium as a coolant, is part of the HTGR family of reactors and thus a product of a lengthy history of research, notably in Germany and the United States. More recently the design has been promoted and revised by the South African utility Eskom and its affiliates. Westinghouse BNFL is a minority investor. Prototype variations of the PBMR are now operating in China and Japan. Eskom has received administrative approval to build a prototype PBMR in South Africa, but has also been delayed in implementation by judicial rulings regarding the reactor's potential environmental impact. Certification procedures in the US have slowed, but have never been abandoned. At around 165 MWe the PBMR is one of the smallest reactors now proposed for the commercial market. This is considered a marketing advantage because new small reactors require lower capital investments than larger new units. Several PBMRs might be built at a single site as local power demand requires. Small size has been viewed as a regulatory disadvantage because most licensing regulations (at least formerly) required separate licences

Figure 8.5 Artist's impression of the Advanced CANDU Reactor [Atomic Energy Canada Ltd (AECL)]

for each unit at a site. The NRC also does not claim the same familiarity with the design that it has with LWRs. Fuels used in the PBMR would include more highly enriched uranium than is now used in LWR designs. The PBMR design is considered a possible contender for the US Department of Energy's Next Generation Nuclear Plant (NGNP) programme in Idaho. China has also indicated interest in building its own variation of the PBMR. China and South Africa have also discussed cooperation in their efforts.

8.6.3 4S

The 4S is a very small molten sodium-cooled reactor designed by Toshiba. The reactor presently being considered is 10 MWe though larger and smaller versions exist. The 4S is designed for use in remote locations and to operate for decades without refuelling. This has led the reactor to be compared with a nuclear 'battery'. The use of molten sodium as a coolant is not particularly new, having been used in many FBR designs. Sodium coolants allow for higher reactor temperatures. Potential fuels are uranium or uranium–plutonium alloys. Uranium is the likely fuel in the United States and present plans call for 19.9 per cent fuel enrichment. This high level of enrichment is one reason the reactor could be able to operate for extended periods without refuelling.

8.6.4 Gas turbine-modular helium reactor

The Gas Turbine-Modular Helium Reactor (GT-MHR) is an HTGR design developed primarily by the US firm, General Atomic. The most advanced plans for GT-MHR development relate to building reactors in Russia to assist in the disposal of surplus plutonium supplies. Parallel plans for commercial power reactors would use uranium-based fuels enriched to as high as 19.9 per cent ^{235}U content. This would keep the fuel just below the 20 per cent enrichment that defines highly enriched uranium. In initial GT-MHR designs, the conversion of the energy to electricity would involve sending the heated helium coolant directly to a gas turbine. There has been concern regarding untested, though non-nuclear aspects of this generation process. This has led potential sponsors to advocate similar ideas involving less innovative heat transfer mechanisms prior to generating electricity or commercial heat. The US utility, Entergy, has participated in GT-MHR development and promotion and has used the name 'Freedom Reactor' for the design. Because coolant temperatures arising from HTGRs are much higher than from LWRs, the design is viewed as an improved commercial heat source. There has been particular attention paid to the design's potential in the production of hydrogen from water. The GT-MHR is considered a potential contender for the US Department of Energy's NGNP programme (see Chapter 9).

Chapter 9
Future reactor designs

9.1 Generation IV

The nuclear designs under consideration for construction in the UK and elsewhere in the next two decades are, at bottom, very similar to those developed in the early days of the industry. This is not surprising: with a history of around 50 years and reactors that can last for 40 or 50 years, the industry has barely had time for its reactor designs to evolve very far. Some would argue that the nuclear industry settled too quickly on its favoured designs and that other options may still offer better efficiency and safety performance. One argument says that the years of experience and technology development built up, for example, in operating a large number of PWRs have counterbalanced the potential of other designs.

One person who argued that this decision was made too early was Klaus Stadie, previously the deputy director of safety and regulation at the OECD's Nuclear Energy Agency. Writing in 1996 on the 40th anniversary of the start up of the first nuclear power reactor, he pointed out that initially eight reactor designs had been selected for construction and operation under the US Atomic Energy Commission's Power Reactor Demonstrator Programme. This was not a systematic review but 'if the programme had been carried through as planned over a period of 15–20 years, it would have provided reactor designers with sufficient insights to choose the one or two most attractive concepts for large power reactors'.

After more than 50 years, several of the early reactor concepts are being revisited and a systematic approach to evaluating them, along with a number of new designs, is being undertaken. Some may become available as a new generation of nuclear reactors around the middle of the century.

This programme is coordinated in the Generation IV International Forum (GIF), an association of 10 countries formed in 2000 that seeks to develop a new generation of commercial nuclear reactor designs before 2030. The members are Argentina, Brazil, Canada, France, Japan, South Korea, South Africa, Switzerland, the UK and the USA.

The forum carried out an evaluation of over 100 potential nuclear designs identified internationally as likely to meet the forum's objectives of secure, clean and proliferation-resistant forms of energy.

Following peer review and evaluation, GIF members agreed during 2002 to concentrate their efforts and funds on six concept designs whose goal is to become commercially viable between 2015 and 2025. The GIF group, along with the US Department of Energy's Nuclear Energy Research Advisory Committee (NERAC), published *'A Technological Roadmap for Generation IV Nuclear Energy Systems'* in 2002, which summarises plans and designs for these six Generation IV projects. There is thus some leeway between the 2030 target for the GIF programme implementation and the targets for individual concepts. Individual GIF participant nations are free to pursue any individual technology they choose. The US is a major mover in the Generation IV programme, initially pursuing all the designs, but it eventually hopes to identify two or three leading designs. It supported the programme in the 2005 Energy Act with over $1.25 billion in direct grants, along with other grants for associated programmes that are researching 'exotic' fuel cycles (i.e. those that include different ways of reprocessing fuel and reusing the recovered uranium and plutonium, and in some cases other fusion products).

A 'Framework Agreement for International Collaboration on Research and Development of Generation IV Nuclear Energy Systems' was signed on 28 February 2005 and so far Canada, France, Japan, South Korea, Switzerland, the UK, the USA and Euratom have joined the framework agreement. A year later (on 15 February 2006) the first GIF system arrangement – the 'System Arrangement for the International Research and Development of the Sodium Cooled Fast Reactor System' – was signed, by France, Japan and the USA.

The designs in the Generation IV research programme are not necessarily entirely new and untried. Some are based on the sodium-cooled fast reactors that have already been in operation in France, the UK, Russia and Japan (see Chapter 2). The ideas behind others date back to the 1960s (the lead-cooled fast reactor) and even the 1950s, including a molten-salt reactor first proposed to fuel a nuclear-powered aircraft in 1954.

In the Generation IV programme the potential uses of these reactors may be broader than the production of electricity. They may also be sited on an industrial site (and sized very differently from current reactors) so that the 60 per cent or more of heat production that is currently vented into air or otherwise wasted can be used to provide process heat for industry. In some designs the reactor is directly linked to a hydrogen production process, anticipating a shift towards using hydrogen, instead of oil, as an energy transfer medium for vehicles and other uses. Finally, the new reactor designs are intended to 'burn' a broader range of fissile substances so that the problem of disposing of plutonium and other long-lived wastes is reduced as they can be transmuted in the reactor. There are six reactors in the programme at present; the US is expected to focus on three for further development. The following summarises the reactor system descriptions in the Technology Roadmap for Generation IV Nuclear Energy Systems and its assessment of research and development needs and other issues associated with each design.

9.2 Gas-cooled fast reactor

The gas-cooled fast reactor (GFR) uses helium coolant directed to a gas turbine generator to produce electricity. This parallels PBMR and earlier high-temperature gas reactor designs, including the GT-MHR designed in the US. The GFR is a fast-spectrum helium-cooled reactor with a closed fuel cycle. The primary difference from early designs is that the GFR would be a 'fast' or breeder reactor.

One favoured aspect of the design is that it would minimise the production of many undesirable spent fuel waste streams. The reference design size was targeted to be rated at 288 MWe, with a deployment target date of 2025.

Like thermal-spectrum helium-cooled reactors such as the GT-MHR and the PBMR, the high outlet temperature of the helium coolant makes it possible to deliver electricity, hydrogen or process heat with high conversion efficiency. The GFR uses a direct-cycle helium turbine for electricity and can use process heat for thermochemical production of hydrogen.

Through the combination of a fast-neutron spectrum and full recycle of actinides, GFRs minimise the production of long-lived radioactive waste isotopes. The GFR's fast-spectrum also makes it possible to utilise available fissile and fertile materials (including depleted uranium from enrichment plants) two orders of magnitude more efficiently than thermal-spectrum gas reactors with once-through fuel cycles. The GFR reference assumes an integrated, on-site spent fuel treatment and refabrication plant.

Research issues for GFR include:

- Development of materials with superior resistance to fast-neutron fluence under very-high-temperature conditions.
- Development of a high-performance helium turbine for efficient generation of electricity.
- Development of efficient coupling technologies for process heat applications and the GFR's high-temperature nuclear heat.

The GFR has several technology gaps in its primary systems and plant balance that are in common with the GT-MHR. Also, the development of very-high-temperature materials with superior resistance to fast-neutron fluence, and innovative refractory fuel concepts with enhanced fission product retention capability, are of generic interest to other types of reactor, including the VHTR and water-cooled reactors.

Target values of some key parameters such as power density and fuel burnup are sufficient for reasonable performance of a first-generation new fuel technology.

Because these parameters have a direct impact on technical and economic performance, there is a strong incentive for additional performance-phase R&D, with the goal of further upgrading the power density to beyond 100 MWth/m^3 and the fuel burnup.

9.2.1 R&D scope

A conceptual design of an entire GFR prototype system can be developed by 2019.

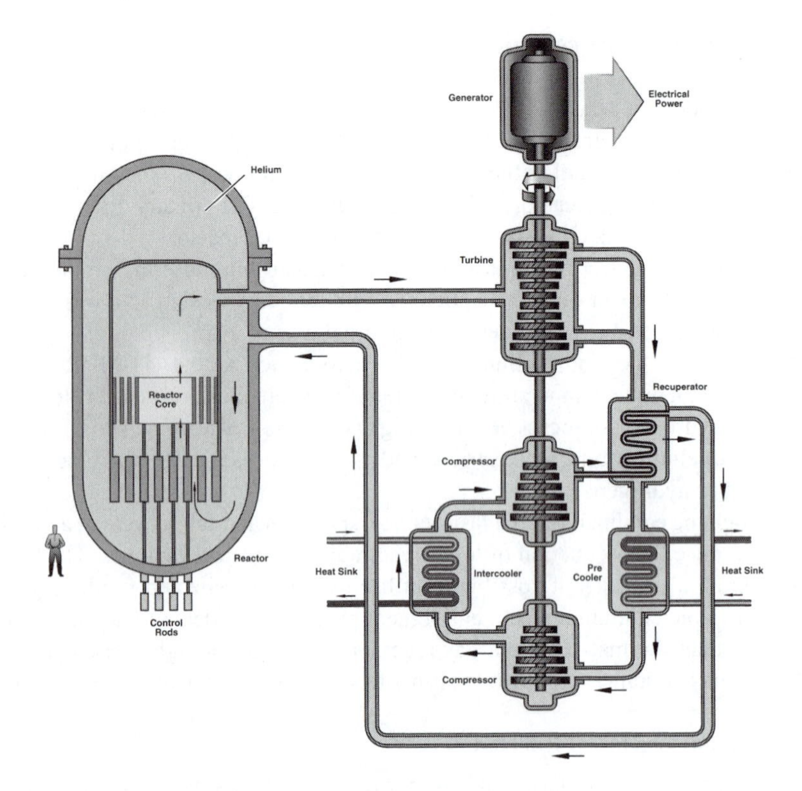

Figure 9.1 Gas-cooled fast reactor (GFR) [Generation IV International Forum]

The prototype system is envisioned as an international project that could be placed in operation by 2025. It would have a reactor power of 600 MW (thermal) and a net plant efficiency as high as 48 per cent by using a direct cycle helium turbine. The fuel compound would be UPuC/SiC (70/30 per cent) with about 20 per cent plutonium content.

9.2.2 Technology base

The technology base for the GFR includes decommissioned reactors such as the Dragon Project, built and operated in the United Kingdom, the AVR and the THTR, built and operated in Germany, and Peach Bottom and Fort St Vrain, built and operated in the United States. Ongoing demonstrations include the HTTR in Japan, which reached full power (30 MWt) using fuel compacts in 1999, and the HTR-10 in China. A consortium of Russian institutes is designing a 300-MWth GT-MHR in cooperation with General Atomics.

The GFR may benefit from development of these technologies, as well as development of innovative fuel and very-high-temperature materials for the VHTR.

Demonstrating the viability of the GFR requires meeting a number of significant technical challenges. Fuel, fuel cycle processes and safety systems pose the major technology gaps.

Candidate fuels. A composite ceramic–ceramic fuel with closely packed, coated (U,Pu)C kernels or fibres is the best option for fuel development. Alternative fuel options for development include fuel particles with large (U,Pu)C kernels and thin coatings or ceramic-clad, solid-solution metal fuels. Fuel fabrication techniques must be developed to be compatible with on-site processing for actinide recovery and remote fuel fabrication. Innovative methods such as vapour deposition or impregnation are among the candidate techniques for on-site manufacturing of composite ceramic fuel (cercer, with cermet as backup).

Candidate materials. The main challenges are in-vessel structural materials, both in-core and out-of-core, that will have to withstand fast-neutron damage and high temperatures, up to 1600 °C in accident situations. Ceramic materials are therefore the reference option for in-core materials, and composite cermet structures or intermetallic compounds will be considered as a backup. For out-of-core structures, metal alloys will be the reference option. For other internal core structures, mainly the upper and lower structures, shielding, the core barrel and grid plate, the gas duct shell and the hot gas duct, the candidate materials are coated or uncoated ferritic–martensitic steels (or austenitic as alternative solution), other Fe–Ni–Cr-base alloys (Inco 800) and Ni-base alloys. The main candidate materials for pressure vessels (reactor, energy conversion system) and cross vessel are 21/4 Cr and 9–12 Cr martensitic steels.

The GFR's innovative design features need to be developed to overcome the shortcomings of previous fast-spectrum gas-cooled designs, which had primarily low thermal inertia and poor heat removal capability at low helium pressure. Various passive approaches will be evaluated for the ultimate removal of decay heat in depressurisation events. The conditions to ensure a sufficient back pressure and to enhance the reliability of flow initiation are some of the key issues for natural convection, the efficiency of which will have to be evaluated for different fuel types, power densities and power conversion units.

Because of the high GFR core power density, a safety approach is required that relies on intrinsic core properties supplemented with additional safety devices and systems as needed but minimises the need for active systems.

9.2.3 Design and evaluation R&D

The most important issues regarding economic viability of the GFR are associated with the simplified and integrated fuel cycle and the modularity of the reactor – this includes volume production, in-factory prefabrication and sharing of on-site resources.

9.2.4 Fuel cycle R&D

The range of fuel options for the GFR underscores the need for early examination of their impacts on the system, especially its fuel cycle. Existing fuel cycle technologies need to be further developed or adapted to allow for the recycling of actinides while preserving the economic competitiveness of the nuclear option in the medium and long term. Laboratory-scale processes for treatment of carbide, nitride or oxide dispersion fuels in ceramic or metal matrices have been evaluated and appear technically feasible. However, extensive experimental work is required in order that the process concepts can be proven feasible for fuel treatment at production scale.

9.2.5 Scale up and demonstration

An important phase of the R&D programme will be to demonstrate, at the level of several kilograms of the selected fuel, the treatment and refabrication of irradiated fuel. The objective is to select and demonstrate the scientific viability of a process by the end of 2012.

9.3 Lead-cooled fast reactor

So far, most breeder reactors have used molten-metal technologies for their coolants. Many FBRs have used molten sodium, a metal with which there is considerable experience but which has sometimes proven difficult to handle. The LFR uses molten lead or a lead–bismuth alloy as its coolant. Similar designs are being investigated in Russia which is not a GIF participant.

Lead-cooled fast reactor systems are lead (Pb) or lead–bismuth (Pb–Bi) alloy-cooled reactors with a fast-neutron spectrum and closed fuel cycle. There is a range of potential plant ratings, including a long refuelling interval battery rated at 50–150 MWe, a modular system of 300–400 MWe and a large monolithic plant at 1,200 MWe. These options also provide a range of energy products.

The battery option is a small factory-built turnkey plant operating on a closed fuel cycle with a very long refuelling interval (15 to 20 years) cassette core or replaceable reactor module. It would be designed to meet market opportunities for electricity production on small grids and for developing countries that may not wish to deploy an indigenous fuel cycle infrastructure to support their nuclear energy systems. Its small size, reduced cost and full support fuel cycle services can be attractive for these markets. It fitted the Generation IV goals best among the LFR options but also had the largest R&D needs and longest development time.

The options in the LFR class may provide a time-phased development path. The nearer-term options focus on electricity production and rely on more easily developed fuel, clad and coolant combinations and their associated fuel recycle and refabrication technologies. The longer-term option seeks to further exploit the properties of lead and raise the coolant outlet temperature sufficiently high to enter markets for hydrogen and process heat, possibly as merchant plants. LFR may advance state-of-the-art liquid-metal fast reactors in the following ways.

- Innovations in heat transport and energy conversion are a central feature of the LFR options. Innovations in heat transport are afforded by natural circulation.
- The favourable properties of lead coolant and nitride fuel, combined with high-temperature structural materials, can extend the reactor coolant outlet temperature

into the 750–800 °C range in the long term, which is potentially suitable for hydrogen manufacture and other process heat applications. In this option, the bismuth alloying agent is eliminated, and since lead is less corrosive it allows the use of new high-temperature materials. R&D would be more extensive than that required for the 550 °C options because the higher reactor outlet temperature requires new structural materials and nitride fuel development.

The technologies employed are extensions of those currently available from the Russian Alpha class submarine Pb–Bi alloy-cooled reactors, from the Integral Fast Reactor metal alloy fuel recycle and refabrication development, and from the ALMR passive safety and modular design approach. Existing ferritic stainless steel and metal alloy fuel, which were developed for sodium fast reactors, are adaptable to Pb–Bi cooled reactors at a reactor outlet temperature of 550 °C.

Innovations in energy conversion are afforded by rising to higher temperatures than liquid sodium allows and by reaching beyond the traditional superheated Rankine steam cycle to supercritical Brayton or Rankine cycles or process heat applications such as hydrogen production and desalination.

The favourable neutronics of Pb and Pb–Bi coolants in the battery option enable low power density, natural circulation-cooled reactors with fissile self-sufficient core designs that hold their reactivity over their very long 15- to 20-year refuelling interval. For modular and large units more conventional higher power density, forced circulation and shorter refuelling intervals are used, but these units benefit from the improved heat transport and energy conversion technology.

Plants with increased inherent safety and a closed fuel cycle can be achieved in the near- to mid-term. The longer-term option is intended for hydrogen production while still retaining the inherent safety features and controllability advantages of a heat transport circuit with large thermal inertia and a coolant that remains at ambient pressure. The favourable sustainability features of fast-spectrum reactors with closed fuel cycles are also retained in all options.

9.3.1 Technology gaps

The important LFR technology gaps are in the areas of

- system fuels and materials, with some gaps remaining for the 550 °C options and large gaps for the 750–800 °C option,
- balance of plant, adapting supercritical steam Rankine or developing supercritical CO_2 electricity production technology,
- economics, focusing on modularisation and factory fabrication,
- fuel cycle technology, including remote fabrication of metal alloy and TRU-N fuels.

9.3.2 Fuels and materials R&D

The nearer-term options use metal alloy fuel or nitride fuel if available. Metal alloy fuel pin performance at 550 °C and metal alloy recycle and refabrication technologies were partly developed for sodium-cooled systems.

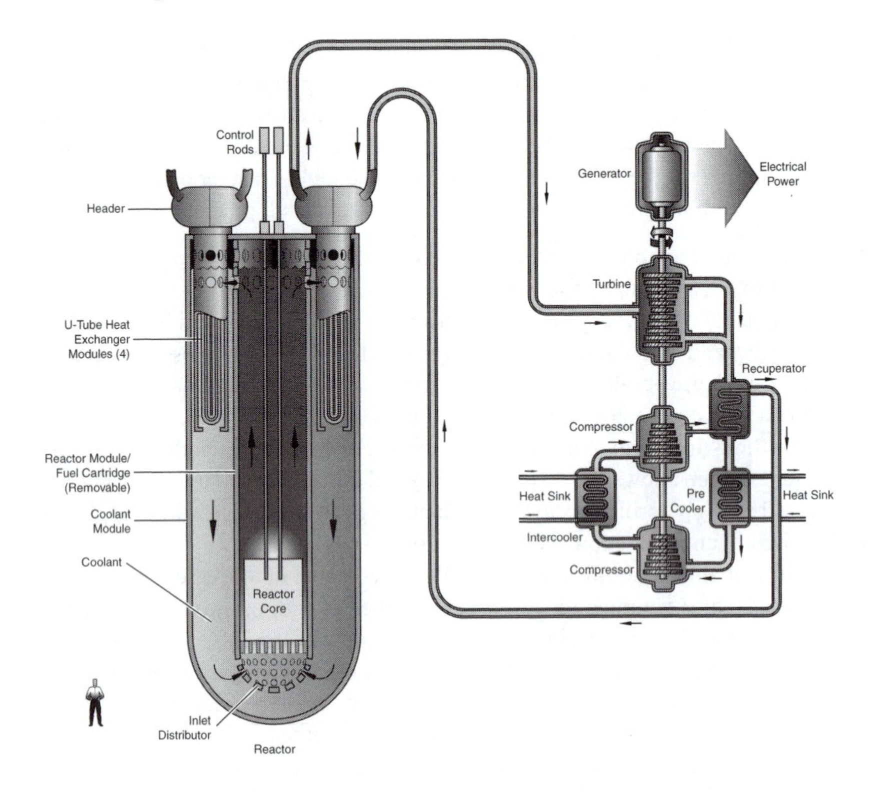

Figure 9.2 Lead-cooled fast reactor (LFR) [Generation IV International Forum]

Mixed nitride fuel is also possible for the 550 °C options; however, it is clearly required for the higher-temperature option. New fuel development will require a long R&D period of 10–15 years to qualify any new fuel for the long-life service conditions in Pb or Pb–Bi.

For process heat applications, an intermediate heat transport loop is needed to isolate the reactor from the energy converter for both safety assurance and product purity.

Viability R&D is also needed for chemistry and activation control of the coolant and corrosion products. Oxygen control is necessary for both Pb and Pb–Bi options. Reactor internals support techniques and refuelling, core positioning and clamping strategies are issues because the internals and the fuel will float (unless restrained) in the dense coolant.

R&D activity is recommended to support the LFR balance of plant in the areas of Ca–Br water cracking for hydrogen production and a supercritical CO_2 Brayton cycle for energy conversion.

9.3.3 Safety R&D

The assurance of reliable and effective thermostructural reactivity feedback is key to the passive safety/passive load following design strategy and will require coordinated neutronics/thermal-hydraulics/structural design of the core. Preliminary testing of mixed nitride fuel under severe upset incore temperature conditions should also be conducted.

Viability R&D activities are needed to determine whether economies can be achieved by plant simplification and reduced footprint, which are afforded by the coolants being inert in air and water, the high conversion efficiency using Brayton cycles or supercritical steam cycles, the economies of mass production, modular assembly and short on-site construction startup time and the production of energy products, possibly including the use of waste heat in a bottoming cycle. Achieving successful economies in the battery and modular options will depend on the adaptation of factory-based mass production techniques from industries such as airplane, truck and auto manufacture and the adaptation of modular/rapid site assembly used for ocean oil rig emplacement and shipbuilding. Life-cycle integrated economics analysis will also be needed that can address modern techniques in design, fabrication, transport, installation and start-up and monitoring and maintenance.

9.3.4 Plant structure

The structural support of the reactor vessel, containing dense Pb or Pb–Bi coolant, will require design development in seismic isolation approaches and sloshing suppression. Also, concrete supports, if used, will have to either be cooled or be designed for high-temperature service.

9.3.5 Fuel cycle

The preferred option for the LFR fuel cycle is pyroprocessing, with advanced aqueous as an alternative. R&D recommended to generally develop the pyroprocess is found in the Crosscutting Fuel Cycle R&D section, although specialisation is required to support the nitride fuel.

Specialisation anticipated for mixed nitride fuel recycle will need to address separations technology, remote refabrication technology, 15N enrichment technologies and irradiation testing. Recycle and remote refabrication R&D activity in the viability phase should involve an iterative screening of conceptual recycle and refabrication approaches, bench-scale testing and flow sheet refinements.

This work will build on existing programmes in Japan and Europe, which are directed to partitioning and transmutation missions. Since 15N enrichment is essential for meeting sustainability goals for waste management (arising from the need to control C-14 production), fuel cycle R&D activity should screen options for 15N enrichment and recovery and associated bench-scale investigations.

9.4 Molten-salt reactor

The molten-salt reactor (MSR) involves a circulating liquid of sodium, zirconium and uranium fluorides as a reactor fuel though the design could use a wide variety of fuel cycles. Versions of the MSR have been around for some time but were never commercially implemented. The MSR was down rated within the Generation IV programme during 2003 because it was seen as too distant in the future for inclusion within the Generation IV schedule. The MSR has been presented as providing a comparatively thorough fuel burn, safe operation and proliferation resistance.

The molten-salt reactor system produces fission power in a circulating molten-salt fuel mixture. MSRs are fuelled with uranium or plutonium fluorides dissolved in a mixture of molten fluorides, with sodium and zirconium fluorides as the primary option. They have good neutron economy, opening alternatives for actinide burning and/or high conversion and operation at high temperature may allow them to be used for thermochemical hydrogen production. The initial reference design would be 1,000 MWe with a deployment target date of 2025. Temperatures would not be as hot as for some other advanced reactors, but some process heat potential exists.

Molten fluoride salts have a very low vapour pressure, so stresses on the vessel and piping are relatively low. Meanwhile, inherent safety is afforded by fail-safe drainage, passive cooling and a low inventory of volatile fission products in the fuel.

Refuelling, processing and fission product removal can be performed online, potentially yielding high availability.

There are four fuel cycle options: using a Th-233 and uranium fuel cycle, using denatured Th-233U converter with minimum inventory of nuclear material suitable for weapons use, a once-through actinide burning (Pu and minor actinides) fuel cycle with minimum chemical processing and actinide burning with continuous recycling. The fourth option with electricity production is favoured for the Generation IV MSR.

Various fluoride salts can be used, some with higher solubility for actinides, others with lower potential for tritium production (if hydrogen production were the objective). Lithium and beryllium fluorides would be preferred if high conversion were the objective. To achieve conversion ratios similar to LWRs, the fuel salt needs only to be replaced every few years. The reactor can use uranium-238 or thorium-232 as a fertile fuel, dissolved as fluorides in the molten salt.

The operating temperatures of MSRs range from the melting point of eutectic fluorine salts (about 450 °C) to below the chemical compatibility temperature of nickel-based alloys (about 800 °C).

9.4.1 Technology base

MSRs were first developed in the late 1940s and 1950s for aircraft propulsion. The aircraft reactor experiment (ARE) in 1954 demonstrated high temperatures (815 °C) and established benchmarks in performance for a circulating fluoride molten salt (NaF/ZrF_4) system.

The 8 MWth molten-salt reactor experiment (MSRE) demonstrated many features, including a lithium/beryllium fluoride salt, graphite moderator, stable performance, off-gas systems and use of different fuels, including uranium-235, uranium-233 and plutonium. A detailed 1,000 MWe engineering conceptual design of a molten-salt reactor was developed. Under these programmes, many issues relating to

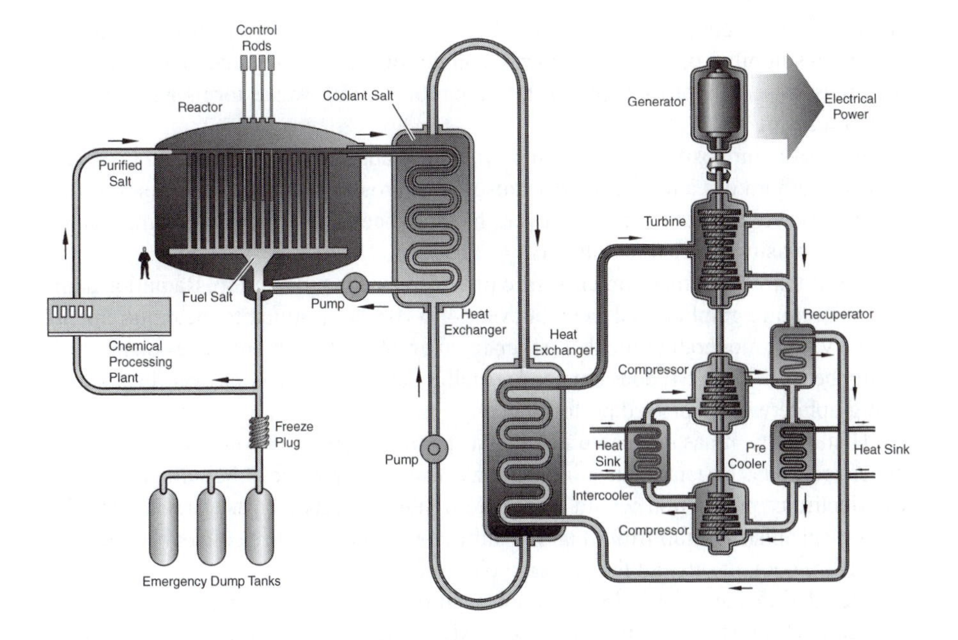

Figure 9.3 Molten-salt reactor (MSR) [Generation IV International Forum]

the operation of MSRs as well as the stability of molten-salt fuel and its compatibility with graphite and Hastelloy N were resolved.

A number of technical viability issues need to be resolved. The highest priority issues include molten-salt chemistry, solubility of actinides and lanthanides in the fuel, compatibility of irradiated molten-salt fuel with structural materials and graphite, and metal clustering in heat exchangers. Specific areas of this viability research phase are as follows:

- Solubility of minor actinides and lanthanides in molten fluoride salt fuel for actinide management with high actinide concentrations.
- Lifetime behaviour of the molten salt fuel chemistry and fuel processing during operation and eventual disposal in a final waste form.
- Materials compatibility with both fresh and irradiated molten-salt fuel for higher-temperature applications.
- Metal clustering (noble metals plate-out on the heat exchanger primary wall).
- Salt processing, separation and reprocessing technology development, including a simplification of the flowsheet.

The main objective of the fuel characterisation research is to develop a simple and reliable chemistry flowsheet that is complete from initial fuel loading to the final waste form. The fuel salt has to meet requirements that include neutronic properties (low neutron cross section for the solvent components, radiation stability, negative temperature coefficient), thermal and transport properties (low melting point, thermal stability, low vapour pressure, adequate heat transfer and viscosity), chemical properties (high solubility of fuel components, compatibility with container and moderator materials, ease of fuel reprocessing), compatibility with waste forms and low fuel and processing costs.

New salt compositions such as sodium and zirconium fluorides should be investigated. Sodium has a higher neutron absorption cross section and is thus somewhat less favourable neutronically. However, this drawback can be partially compensated for by increasing the fuel enrichment.

The graphite's primary function is to provide neutron moderation. Radiation damage will require graphite replacement every 4 to 10 years, similar to the requirements for the VHTR moderator blocks. Longer-lived graphite directly improves plant availability because the MSR does not need refuelling outages. This is a driver for research into graphite with improved performance.

Historically, it has been assumed that a steam power cycle would be used to produce electricity. Recent studies indicate that use of an advanced helium gas turbine for electricity production would increase efficiency, reduce costs, provide an efficient mechanism to trap tritium and avoid potential chemical reactions between the secondary coolant salt and the power cycle fluid.

Detailed design of an MSR has not been done since 1970. An updated design (including design tradeoff studies) is required to better understand strengths and weaknesses and allow defensible economic evaluations. The current regulatory structure is designed for solid-fuel reactors, and the MSR design needs to carefully address the intent of current regulations. Work is required with regulators to define equivalence in safety for MSRs. Because the MSR shares many features with reprocessing plants, the development of MSR regulatory and licensing approaches should be coordinated with R&D in pyroprocessing. Under the high-radiation and -temperature environment, remote and robotic maintenance, inspection and repair are key technologies that require R&D.

9.5 Sodium-cooled fast reactor

Sodium-cooled fast reactors (SFRs) have been the most popular design for breeder reactors. Designs have been proposed under the Department of Energy's 'roadmap' for Generation IV reactors ranging from 150 to 1,700 MWe.

Molten-metal technology is no longer 'new' but several early SFR prototypes had difficulty obtaining sustained operation. The BN-600 in Russia, however, has been regarded as highly reliable. Design supporters believe that the SFR promises superior fuel management characteristics. The original target deployment date of 2015 reflected the considerable research that the design has already received though the design is clearly not as ready for US deployment as LWR designs being evaluated for roughly the same period. The target date seems to be lagging as the VHTR designs gain favour.

The SFR system features a fast-spectrum reactor and closed fuel recycle system. The primary mission for the SFR is management of high-level wastes and, in particular, management of plutonium and other actinides. With innovations to reduce capital cost, the mission can extend to electricity production, given the proven capability of sodium reactors to utilise almost all of the energy in the natural uranium versus the 1 per cent utilised in thermal spectrum systems.

A range of plant size options is available for the SFR, ranging from modular systems of a few hundred megawatt electric to large monolithic reactors of 1,500–1,700 MWe. Sodium-core outlet temperatures are typically 530–550 °C.

The primary coolant system can either be arranged in a pool layout (a common approach, where all primary system components are housed in a single vessel) or in a compact loop layout, favoured in Japan. For both options, there is a relatively large thermal inertia of the primary coolant. A large margin to coolant boiling is achieved by design and is an important safety feature of these systems. Another major safety feature is that the primary system operates at essentially atmospheric pressure, pressurised only to the extent needed to move fluid.

Sodium reacts chemically with air, and with water, and thus the design must limit the potential for such reactions and their consequences. To improve safety, a secondary sodium system acts as a buffer between the radioactive sodium in the primary system and the steam or water that is contained in the conventional Rankine-cycle power plant. If a sodium–water reaction occurs, it does not involve a radioactive release. Still, there is important viability work to be done in safety. Key needs are to confirm reliability of passive feedback from heatup of reactor structures and to establish the long-term coolability of oxide or metal fuel debris after a bounding case accident.

There are two fuel options: MOX and mixed uranium–plutonium–zirconium metal alloy (metal). The experience with MOX fuel is considerably more extensive than with metal. SFRs require a closed fuel cycle to enable their advantageous actinide management and fuel utilisation features. The reactor technology and the fuel cycle technology are strongly linked. Consequently, much of the research recommended for the SFR is relevant to crosscutting fuel cycle issues.

The options for fuel recycle are the advanced aqueous process and the pyroprocess. The technology base for the advanced aqueous process comes from long experience in several countries with PUREX process technology. The advanced process proposed by Japan, for example, is simplified relative to PUREX and does not result in highly purified products. The technology base for fabrication of oxide fuel assemblies is substantial, yet further extension is needed to make the process remotely operable and maintainable.

The important technology gaps for the SFR are in the following areas.

- Ensuring a passively safe response to all design basis initiators, including anticipated transients without scram (a major advantage for these systems),
- Capital cost reduction,
- Proof by test of the reactor's ability to accommodate bounding events.

9.5.1 Technology base

Sodium-cooled liquid-metal reactors are the most technologically developed of the six Generation IV systems. SFRs have been built and operated in France, Japan,

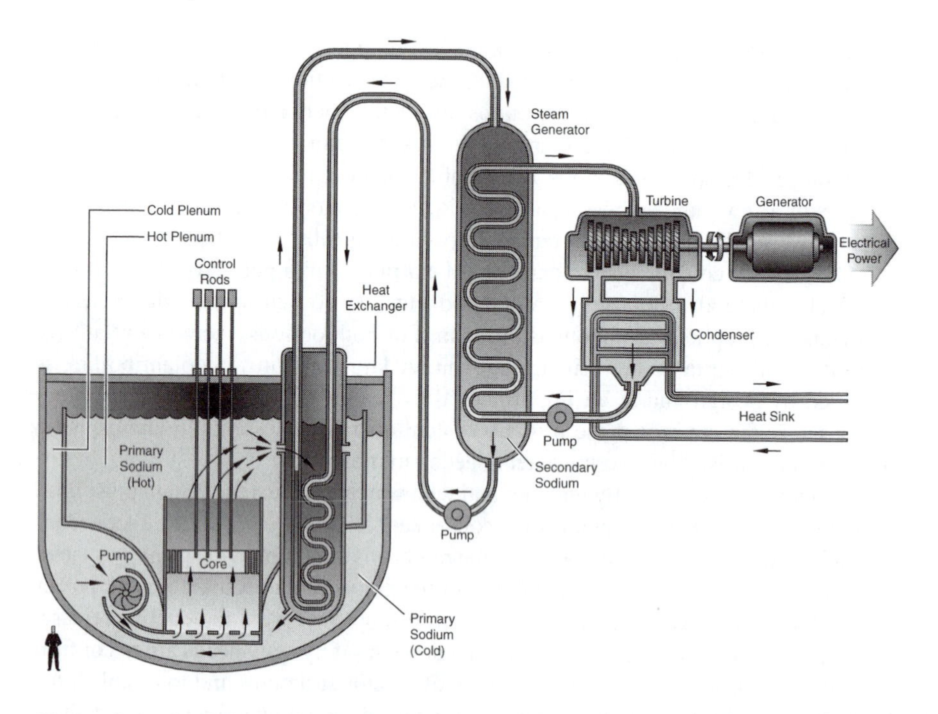

Figure 9.4   Sodium-cooled fast reactor (SFR) [Generation IV International Forum]

Germany, the United Kingdom, Russia and the USA. Demonstration plants ranged from 1.1 MWt (at EBR-I in 1951) to 1,200 MWe (at SuperPhénix in 1985), and sodium-cooled reactors are operating today in Japan, France and Russia. As a benefit of these previous investments in technology, the majority of the R&D needs are performance-related. With the exception of passive safety assurance, there are few viability issues with regard to the reactor systems.

Other important SFR reactor technology gaps are in-service inspection and repair (in sodium) and completion of the fuels database.

A key performance issue for the SFR is cost reduction to competitive levels. The extent of the technology base for SFRs is noted above, yet none of the SFRs constructed to date has been economical to build or operate. However, design studies have been done, some of them very extensively, in which proponents conclude that both overnight cost and busbar cost can be comparable to or lower than those of the advanced LWRs. Ultimately, cost reductions are best if supported by specific innovations, providing a better measure of confidence.

9.5.2   Reactor systems R&D

The reactor technology R&D is aimed at enhancing the economic competitiveness and plant availability. For example, development and/or selection of structural materials for components and piping is important for the development of an economically competitive plant. In place of austenitic steels, ferritic steels (containing 12 per cent

Cr) are viewed as promising structural materials for future plant components because of their superior elevated temperature strength and thermal properties.

Improvement of in-service inspection and repair technologies is important to confirm the integrity of safety-related structures and boundaries that are submerged in sodium and to repair them in place. Motivated by the need to address sodium–water reactions, it is also important to enhance the reliability of early detection systems for water leaks. New early detection systems, especially those that protect against small leaks, would be adopted to prevent the propagation of tube ruptures and to allow a rapid return to plant operation. Noting the temperatures at which the SFRs operate, there may be interest in investigating the use of a supercritical CO_2 Brayton cycle.

A focused programme of safety R&D is necessary to support the SFR. The safety R&D challenges for these systems in the Generation IV context are to verify the predictability and effectiveness of the mechanisms that contribute to passively safe response to design basis transients and anticipated transients without scram and to ensure that bounding events considered in licensing can be sustained without loss of coolability of fuel or loss of containment function.

Since many of the mechanisms that are relied upon for passively safe response can be predicted on a first-principles basis (e.g. thermal expansion of the fuel and core grid plate structure), enough is now known to perform a conceptual design of a prototype reactor. R&D is recommended to evaluate physical phenomena and design features that can be important contributors to passive safety and to establish coolability of fuel assemblies if damage should occur. This R&D would involve in-pile experiments, primarily on metal fuels, using a transient test facility. While there are design studies in progress in Japan on SFRs, there is little design work in the United States, even at the preconceptual level. Design work is an important performance issue, and it should accelerate given the importance of economics for the SFR. R&D activity is needed with a focus on the base technology for component development.

9.6 Supercritical-water-cooled reactor

Supercritical-water-cooled reactor (SCWR) is to be the next step in LWR development and has been proposed with alternatives that evolve from both the BWR and the PWR. SCWRs would operate at higher temperatures and thermal efficiencies than present LWRs. The reference plant might be 1,700 MWe – the upper end of present LWR designs. The deployment target date would be 2025. Some GIF participants favour the SCWR design because it is more familiar to commercial markets than are more innovative concepts. Much of the design research has been done in Japan. SCWR systems are high-temperature, high-pressure water-cooled reactors that operate above the thermodynamic critical point of water (374 °C, 22.1 MPa).

These systems may have a thermal or fast-neutron spectrum, depending on the core design. SCWRs have unique features that may offer advantages compared with state-of-the-art LWRs, as given below.

- Increased thermal efficiency relative to current-generation LWRs. The efficiency of an SCWR can approach 44 per cent, compared to 33–35 per cent for LWRs.

- A lower-coolant mass flow rate per unit core thermal power results from the higher enthalpy content of the coolant. This offers a reduction in the size of the reactor coolant pumps, piping and associated equipment and a reduction in the pumping power.
- A lower-coolant mass inventory results from the once-through coolant path in the reactor vessel and the lower-coolant density. This opens the possibility of smaller containment buildings.
- No boiling crisis exists due to the lack of a second phase in the reactor, thereby avoiding discontinuous heat transfer regimes within the core during normal operation.
- Steam dryers, steam separators, recirculation pumps and steam generators are eliminated. Therefore, the SCWR can be a simpler plant with fewer major components.

The Japanese supercritical-light-water reactor (SCLWR) with a thermal spectrum has been the subject of most development work in the last 10 to 15 years and is the basis for much of the reference design.

The SCLWR reactor vessel is similar in design to a PWR vessel (although the primary coolant system is a direct-cycle, BWR-type system). High-pressure (25.0 MPa) coolant enters the vessel at 280 °C. The inlet flow splits, partly to a downcomer and partly to a plenum at the top of the core to flow down through the core in special water rods. This strategy provides moderation in the core. The coolant is heated to about 510 °C and delivered to a power conversion cycle, which blends LWR and supercritical fossil plant technology.

The overnight capital cost for a 1,700-MWe SCLWR plant may be as low as $900/kWe (about half that of current ALWR capital costs), considering the effects of simplification, compactness and economy of scale. The operating costs may be 35 per cent less than current LWRs. The SCWR can also be designed to operate as a fast reactor. The difference between thermal and fast versions is primarily the amount of moderator material in the SCWR core. The fast-spectrum reactors use no additional moderator material, while the thermal spectrum reactors need additional moderator material in the core.

9.6.1 Technology base

Much of the technology base for the SCWR can be found existing in LWRs and in commercial supercritical-water-cooled fossil-fired power plants. However, there are some relatively immature areas. There have been no prototype SCWRs built and tested. For the reactor primary system, there has been very little in-pile research done on potential SCWR materials or designs, although some SCWR in-pile research has been done for defence programmes in Russia and the USA. Limited design analysis has been underway over the last 10 to 15 years in Japan, Canada and Russia. For the balance of plant, there has been development of turbine generators, piping and other equipment extensively used in SCW-cooled fossil-fired power plants.

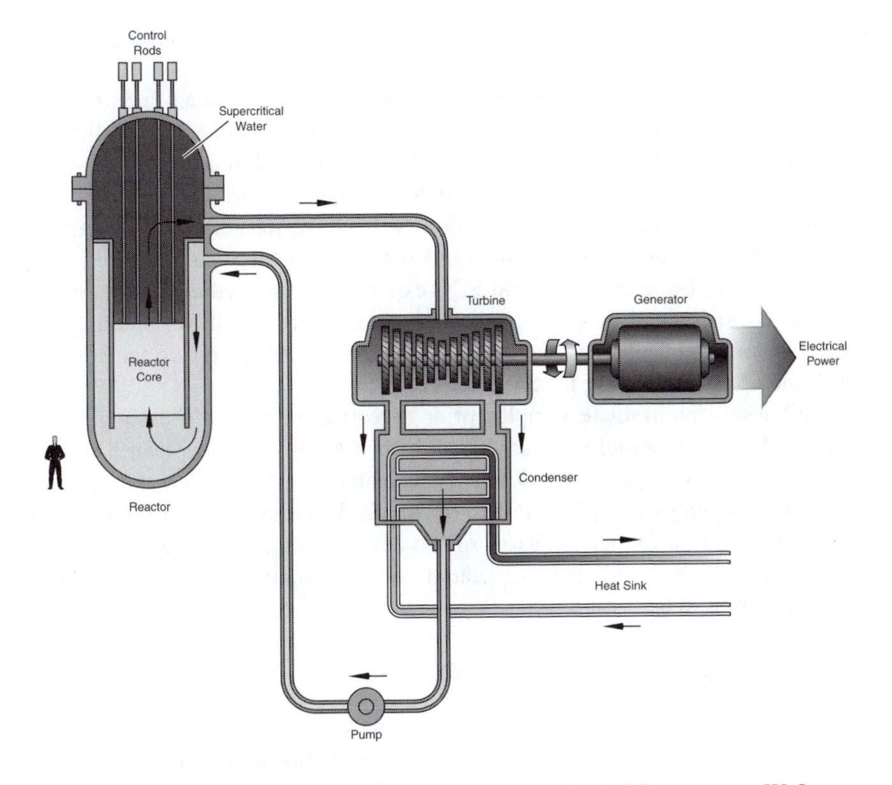

Figure 9.5 Supercritical-water-cooled reactor (SCWR) [Generation IV International Forum]

The important SCW technology gaps are in the areas of the following:

- materials and structures
- safety, including power-flow stability during operation
- plant design.

Important viability issues are found within the first two areas, and performance issues are found primarily within the first and third areas.

The SCW environment is unique and few data exist on the behaviour of materials in SCW under irradiation and in the temperature and pressure ranges of interest. At present, no candidate alloy has been confirmed for use as either the cladding or structural material in thermal or fast-spectrum SCWRs.

Potential candidates include austenitic stainless steels, solid solution and precipitation-hardened alloys, ferritic–martensitic alloys and oxide dispersion-strengthened alloys. The fast SCWR design would result in greater doses to cladding and structural materials than in the thermal design by a factor of five or more. These doses will result in greater demands on the structural materials in terms of the need for irradiation stability and effects of irradiation on embrittlement, creep, corrosion and SCC.

The generation of helium by transmutation of nickel is also an important consideration in both the thermal and fast designs because it can lead to swelling and embrittlement at high temperatures. The data obtained during prior fast reactor development will play an important role in this area.

The corrosion and SCC R&D activities will be organised into three parts: an extensive series of out-of-pile corrosion and SCC experiments on unirradiated alloys, companion out-of-pile corrosion and SCC experiments on irradiated alloys and in-pile loop corrosion and SCC tests. It is envisioned that at least two and maybe as many as four out-of-pile test loops would be used, some addressing the corrosion issues and others addressing the SCC issues.

Facilities to preirradiate samples prior to corrosion and SCC testing will be required. This work should be carried out over a 6–10 year time span for unirradiated materials and the same for irradiated materials. About 10 years of in-pile testing in these loops will be needed to obtain all the required data to support both the viability and performance phases of the development of the thermal-spectrum version of the SCWR and about 15 years to obtain the needed information for the fast-spectrum SCWR.

9.6.2 Reactor systems R&D

A number of reactor system alternatives have been developed for both vessel and pressure tube versions of the SCWR. On the conventional side, it can utilise the existing technology from the secondary side of the supercritical-water-cooled fossil-fired plants. Significant research in this area is not needed.

An SCWR safety research activity is recommended, organised around the following topics.

- Reduced uncertainty in SCW transport properties
- Further development of appropriate fuel cladding to coolant heat transfer correlations for SCWRs under a range of fuel rod geometries
- SCW critical flow measurements, as well as models and correlations
- Measurement of integral loss of coolant accident (LOCA) thermal-hydraulic phenomena
- Fuel rod cladding ballooning during LOCAs
- SCWR design optimisation studies, including investigations to establish the reliability and system cost impacts of passive safety systems
- Power-flow stability assessments.

9.6.3 Design and evaluation

Many of the major systems that can potentially be used in an SCWR were developed for the current BWRs, PWRs and SCW fossil plants. Therefore, the major plant design and development needs that are unique for SCWRs are primarily found in their design optimisation, as well as their performance and reliability assurance under SCWR neutronic and thermo-hydraulic conditions. Two major differences in conditions are the stresses due to the high SCWR operating pressure (25 MPa) and the large coolant temperature and density change (approximately 280 to 500 °C or more, 800 to 80 kg/m^3, respectively) along the core under the radiation field. Examples of design features that need to be optimised to achieve competitiveness in economics without sacrificing safety or reliability include the fuel assemblies, control rod drive system, internals, reactor vessel, pressure relief values, coolant clean-up system, reactor control logic, turbine configuration, re-heaters, de-aerator, start-up system and procedures, in-core sensors and containment building. This work is expected to take about 8 to 10 years.

9.7 Very-high-temperature reactor

Very-high-temperature reactor (VHTR) is an evolution from the HTGR family of reactors but would operate at even higher temperatures than designs now undergoing pre-certification. Some of the VHTR design standards might be met by modified PBMRs or GT-MHRs.

In contrast to the GFR, the VHTR would not be a breeder reactor, thus it would produce less potentially usable fuel than it consumes. In addition to generating electricity, the design can provide process heat for industrial activities including hydrogen production and desalinisation. Deployment is targeted for 2020, earlier than most Generation IV designs. The VHTR is now a favoured design in the US, where it is the basis for most anticipated submissions for the still-evolving next generation nuclear plant (NGNP).

The VHTR system is a next step in the evolutionary development of high-temperature gas-cooled reactors. The VHTR can produce hydrogen from only heat and water by using the thermochemical iodine–sulphur (I–S) process or from heat, water and natural gas by applying the steam reformer technology to core outlet temperatures greater than about 1,000 °C.

A 600 MWth VHTR dedicated to hydrogen production could yield over 2 million cubic metres per day. The VHTR can also generate electricity with high efficiency, over 50 per cent at 1,000 °C, compared with 47 per cent at 850 °C in the GT-MHR or PBMR.

Cogeneration of heat and power makes the VHTR an attractive heat source for large industrial complexes. The VHTR can be deployed in refineries and petrochemical industries to substitute large amounts of process heat at different temperatures, including hydrogen generation for upgrading heavy and sour crude oil. Core outlet temperatures higher than 1,000 °C would enable nuclear heat application to such processes as steel, aluminium oxide and aluminium production.

The VHTR is a graphite-moderated, helium-cooled reactor with thermal neutron spectrum. It can supply nuclear heat with core outlet temperatures of 1,000 °C. The reactor core type of the VHTR can be a prismatic block core such as the operating Japanese HTTR or a pebble-bed core such as the Chinese HTR-10. For electricity generation, the helium gas turbine system can be directly set in the primary coolant

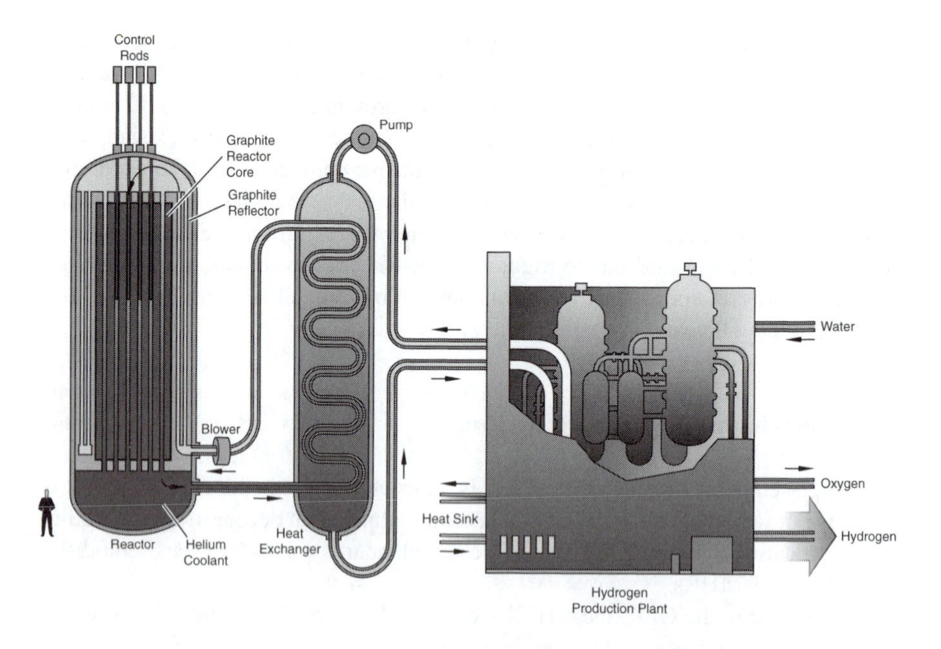

Figure 9.6 Very-high-temperature reactor (VHTR) [Generation IV International Forum]

loop, which is called a direct cycle. For nuclear heat applications, such as process heat for refineries, petrochemistry, metallurgy and hydrogen production, the heat application process is generally coupled with the reactor through an intermediate heat exchanger (IHX), which is called an indirect cycle.

9.7.1 Technology base

The VHTR evolves from HTGR experience and extensive international databases that can support its development. The basic technology for the VHTR has been well established in former HTGR plants, such as Dragon, Peach Bottom, AVR, THTR and Fort St Vrain and is being advanced in concepts such as the GT-MHR and PBMR. The ongoing 30-MWth HTTR project in Japan is intended to demonstrate the feasibility of reaching outlet temperatures up to 950 °C coupled to a heat utilisation process, and the HTR-10 in China will demonstrate electricity and cogeneration at a power level of 10 MWth. The former projects in Germany and Japan provide data relevant to VHTR development. Steam reforming is the current hydrogen production technology. The coupling of this technology will be demonstrated in large scale in the HTTR programme but still needs complementary R&D for market introduction.

9.7.2 Technology gaps

Process-specific R&D gaps exist to adapt the chemical process and the nuclear heat source to each other with regard to temperatures, power levels and operational pressures. Heating of chemical reactors by helium is different from current industrial practice and needs specific R&D and demonstration. Qualification of high-temperature alloys and coatings for resistance to corrosive gases such as hydrogen, carbon monoxide and methane will be needed.

The viability of producing hydrogen using the iodine–sulphur (I–S) process still requires pilot- and large-scale demonstration of the three basic chemical reactions and development of corrosion-resistant materials. Any contamination of the product will have to be avoided. Development of heat exchangers, coolant gas ducts and valves will be necessary for isolation of the nuclear island from the production facilities. This is especially the case for isotopes like tritium, which can easily permeate metallic barriers at high temperatures.

Performance issues for the VHTR include development of a high-performance helium turbine for efficient generation of electricity. Modularisation of the reactor and heat utilisation systems is another challenge for commercial deployment of the VHTR.

The increase in the helium core outlet temperature of the VHTR results in an increase in the fuel temperature and reduced margins in case of core heatup accidents. Fuel particles coated with silicon-carbide are used in HTGRs at fuel temperatures of about 1,200 °C. Irradiation testing is required to demonstrate that TRISO-coated particles can perform acceptably at the high burnup and temperature associated with the VHTR. Following irradiation, high-temperature heating (safety) tests are needed to determine that there is no degradation in fuel performance under accident heatup conditions up to 1,600 °C as a result of the more demanding irradiation service conditions. These fuel demonstration activities would require about 5 to 7 years to complete following fabrication of samples. Demonstrating the viability of the VHTR core requires meeting a number of significant technical challenges. Novel fuels and materials must be developed that:

- Permit increasing the core outlet temperatures from 850 °C to 1,000 °C and preferably higher.
- Permit the maximum fuel temperature reached following accidents to reach 1,800 °C.
- Permit maximum fuel burnup of 150–200 GWD/MTHM.
- Avoid power peaking and temperature gradients in the core, as well as hot streaks in the coolant gas.

Only laboratory-scale fabrication of ZrC-coated particle fuel has been performed to date. Research into more economical commercial-scale fabrication routes for ZrC-coated particle fuels, including process development at production scale, is required. Advanced coating techniques or advanced processing techniques (automation) should be considered.

Increasing the allowable fuel burnup requires development of burnable absorbers for reactivity control. The behaviour of burnable absorbers needs to be established (irradiation dimensional stability, swelling, lifetime) under the design service conditions of the VHTR. Development of carbon–carbon composites is needed for control rod sheaths, especially for the VHTR based on a prismatic block core, so that the control rods can be inserted into the high-temperature areas entirely down to the core.

Promising ceramics such as fibre-reinforced ceramics, sintered alpha silicon-carbide, oxide-composite ceramics, and other compound materials are also being developed for other industrial applications needing high-strength, high-temperature materials. The feasibility of using superplastic ceramics in VHTR components will be investigated by studying the effects of neutron irradiation on superplastic deformation mechanisms.

Testing of core internals is envisioned to take 5 to 10 years at any of the test reactors worldwide.

To realise the goal of core outlet temperatures higher than 1,000 °C, new metallic alloys for reactor pressure vessels have to be developed. At these core outlet temperatures, the reactor pressure vessel temperature will exceed 450 °C, but LWR pressure vessels were developed for 300 °C service and the HTTR vessel for 400 °C. Hasteloy-XR metallic materials are used for intermediate heat exchanger and high-temperature gas ducts in the HTTR at core outlet temperatures up to about 950 °C, but further development of Ni–Cr–W super-alloys and other promising metallic alloys will be required for the VHTR.

The irradiation behaviour of these super-alloys under the service conditions expected in the VHTR will need to be characterised. Such work is expected to take 8 to 12 years.

An alternative pressure vessel allowing for larger diameters and ease of transportation, construction and dismantling would be the pre-stressed cast-iron vessel, which can also prevent a sudden burst due to separation of mechanical strength and leak tightness. The vessel could also include a passive decay heat removal system with enhanced efficiency. Internal core structures and cooling systems, such as intermediate heat exchanger, hot gas duct, process components and isolation valve, that are in contact with the hot helium can use the current metallic materials up to about 1,000 °C core outlet temperature. For core outlet temperatures exceeding 1,000 °C, ceramic materials must be developed. Piping and component insulation also requires design and materials development.

Core internal structures containing the fuel elements such as pebbles or blocks are made of high-quality graphite. The performance of high-quality graphite for core internals has been demonstrated in gas-cooled pilot and demonstration plants, but recent improvements in the manufacturing process of industrial graphite have shown improved oxidation resistance and better structural strength. Irradiation tests are needed to qualify components using advanced graphite or composites to the fast fluence limits of the VHTR.

The balance of a VHTR plant is determined by the specific application, which can be thermochemical processes, dedicated electricity production or cogeneration. All components have to be developed for temperatures well above the present state-of-the-art and depend on a comprehensive materials qualification activity. Failure mechanisms such as creep, fretting and ratcheting have to be studied in detail, precluded with design, and demonstrated in component tests. Specific components such as IHX, isolation valves, hot gas ducts with low heat loss, steam reformers and process-related heat exchangers have to be developed for use in the modular VHTR, which mainly uses only one loop. This leads to much larger components than formerly developed and a new design approach by modularisation of the component itself. Low pressures are necessary or preferable for many processes.

Alternative coolants for the intermediate loop such as molten salt should be adapted where needed.

Other applications will require different components such as helium-heated steam crackers, distiller columns and superheaters.

Passive heat removal systems should be developed to facilitate operation of the VHTR and analysis and demonstration of its inherent safety features are needed. Additional safety analysis is necessary with regard to nuclear process heat applications in an industrial environment.

Design basis and severe accident analyses for the VHTR will need to include phenomena such as chemical attack of graphitic core materials, typically either by air or water ingress.

9.7.3 Fuel cycle

The VHTR assumes a once-through, LEU (<20 per cent ^{235}U) fuel cycle. Like LWR spent fuel, VHTR spent fuel could be disposed of in a geologic repository or conditioned for optimum waste disposal. The current HTGR particle fuel coatings form an encapsulation for the spent fuel fission products that is extremely resistant to leaching in a final repository. However, as removed from the reactor, the fuel includes large quantities of graphite, and research is required to define the optimum packaging form of spent VHTR fuels for long-term disposal. Radiation damage will require graphite replacement every 4 to 10 years. An optimised approach for dealing with the graphite is necessary.

Recycling of LWR and VHTR spent fuel in a symbiotic fuel cycle can achieve significant reductions in waste quantities and radiotoxicity because of the VHTR's ability to accommodate a wide variety of mixtures of fissile and fertile materials without significant modification of the core design. This flexibility was demonstrated in the AVR test reactor in Germany and is a result of the ability of gas reactors to decouple the optimisation of the core cooling geometry from the neutronics. For an actinide burning alternative, specific Pu-based driver fuel and transmutation fuel containing minor actinides would have to be developed. This fuel can benefit from the abovementioned R&D on SiC and ZrC coating but will need more R&D than LEU fuel.

Chapter 10
The development of fusion

10.1 The theory of fusion

The work of Albert Einstein had shown that if two nuclei fused there would be an energy release, but classical physics says that two particles with the same sign of electrical charge will repel each other, like two opposed magnets. There are clearly exceptions at the atomic scale. In nuclear fission reactors, the particle that causes the fission reaction – the neutron – has no charge, so it is not repelled by the nucleus and its positively charged protons. This is not the case with, for example, the alpha article, which has a double positive charge, or with some other forms of radiation.

In 1928, in parallel with the developing work on the structure of the atom and nucleus (see Chapter 1) George Gamow, a Russian-American theoretical physicist, derived a quantum-mechanical formula which allowed that in some cases two charged particles could overcome their mutual electrostatic repulsion and come very close together. This quantum-mechanical probability is now known as the 'Gamow factor'. It is widely used to explain the measured rates of certain radioactive decays. It was necessary to know the Gamow factor to estimate how often two nuclei with the same sign of electrical charge would get close enough together to fuse and thereby generate energy.

Meanwhile, theoretical explanations were being developed for energy production in the sun and other stars and the fusion of atomic nuclei was one possibility. In the decade that followed Gamow's work, he and other physicists used the Gamow factor to derive the rate at which nuclear reactions would proceed at the high temperatures and pressures believed to exist in the interiors of stars.

In 1938 Hans Bethe worked out the basic nuclear processes by which hydrogen is 'burned' (i.e. completes fusion with other particles) into helium in stellar interiors. Hydrogen is the most abundant constituent of the sun and similar stars, and indeed the most abundant element in the universe. Bethe described the results of his calculations in a paper entitled 'Energy Production in Stars'. He analysed the different possibilities for reactions that fuse nuclei and selected as most important the two processes that we now believe are responsible for sunshine. One process, the so-called p—p chain, builds helium out of hydrogen and is the dominant energy source in stars like the sun and less massive stars. Bethe was rewarded for his discoveries in 1967 with the Nobel Prize for Physics.

Bethe proposed that in the huge temperatures at the centre of the sun – which reaches temperatures of up to 10–15 million degrees and is caused by the gravitational pull of the huge mass of matter – protons would be fused and, after two or three steps involving six protons, would have formed helium and in addition would have released two protons to continue the fusion process. Other physicists later refined this process and suggested other fusion reactions.

The first successful fusion reactions on earth were produced in weapons, where the enormous temperatures and pressures required could be achieved. But work on developing a controlled fusion reaction to provide power and energy was also under way.

In the UK much of the early work on fusion was undertaken by universities, before being centred at Harwell and Aldermaston. The original large-scale experimental fusion device on which British physicists worked during the 1940s and 1950s was housed in a hangar at Harwell and was known as the Zero Energy Toroidal Assembly (ZETA).

The ZETA project was at first shrouded in secrecy, but when there was a temporary thaw in the Cold War, created in the late 1950s by the visit of Kruschev and Bulganin, the work was opened up. The Russians brought their leading fusion expert, Academician IV Kurchatov, to give a lecture on 'The Possibility of Producing Thermonuclear Reactions in a Gas Discharge'. This revealed the Soviets' own work in the field of fusion and the UK shared its experience with ZETA. In fact, according to one source, in the 1950s the work on producing a practical power reactor using nuclear fusion was proceeding so slowly that all the work done on the problem – although not on its use in bombs – was declassified in 1958. The consequent declassification led directly to the setting up of a custom-built laboratory at Culham.

International cooperation began on the fusion power reactor itself, while the subsidiary processes and components required became part of mainstream physics and engineering research and development. The international nature of the power development has been a prerequisite in the development of fusion research given the long timescales and high costs involved.

Most of the world participates in fusion research to a greater or lesser extent, with the principal countries involved in large-scale fusion research being the European Union, USA, Russia and Japan, supported by vigorous programmes in China, Brazil, Canada and Korea.

10.2 The goal

Is the pursuit of fusion necessary? There are a number of other factors which must be taken into consideration. In 1990 some 75 per cent of the world's population (those in the developing countries) were responsible for only 33 per cent of the world's energy consumption; by the year 2020 that 75 per cent is likely to have risen to 85 per cent and the energy consumption to around 55 per cent. Thus there will be greater competition for the fuel resources available.

Another important factor is likely to be a further tightening of international agreements regarding CO_2 emissions to decelerate the effects of global warming and consequent climatic changes. All this amounts to the need for intensified scientific research to achieve greater efficiency and conservation of our energy resources.

Fusion offers significant potential advantages as a future source of energy as part of a varied world energy mix.

First, the necessary fuel is abundant. The reaction is fuelled by deuterium (a heavy form of hydrogen with one neutron added to the usual single proton in the nucleus), which can be extracted from all water bodies. The hydrogen in water molecules is most frequently found as 'semi-heavy' water, in which one of the two hydrogen atoms is replaced with deuterium. The proportion of semi-heavy water molecules is around 1 in 3,200 and it can be separated by distillation or electrolysis. If all the world's current electricity use were to be provided by fusion power stations, present deuterium supplies from water would last millions of years.

The second fuel component is tritium (a superheavy form of hydrogen with one proton and two neutrons in the nucleus). Tritium does not occur naturally and will be bred from lithium within the machine. Therefore, once the reaction is established, even though it occurs between deuterium and tritium, the external fuels required are deuterium and lithium. Lithium is the lightest metallic element and is plentiful in the earth's crust. If all the world's current electricity needs were to be provided by fusion, known lithium reserves would last at least one thousand years.

The energy gained from a fusion reaction is enormous. To illustrate, 10 g of deuterium (which can be extracted from 500 l of water) and 15 g of tritium (produced from 30 g of lithium) reacting in a fusion power plant would produce enough energy for the lifetime electricity needs of an average person in an industrialised country.

The imperative to demonstrate that fusion has the potential to be a safe and clean method of generating base load electricity led to the setting up of the European Safety and Environmental Assessment of Fusion Power (SEAFP) team in 1992. The main participants in SEAFP were the NET (Next Experimental Torus) team, the UKAEA, other European fusion laboratories and a grouping of major European industrial companies.

The work embraced the conceptual design of fusion power stations and the safety and environmental assessments of those designs. Detailed work was done on the identification and modelling of conceivable accident sequences, the potential hazards of normal operation, waste management, the long-term availability of materials and other issues.

The major conclusions reached by the SEAFP team in 1995 were that fusion has very good inherent safety qualities; there are no chain reactions and no production of actinides. The worst possible accident originating in a fusion power station could not breach the confinement; any releases could not approach levels at which evacuation would be considered.

The amount of deuterium and tritium present in the fusion reactor and undergoing fusion at any one time is very small (just a few grammes) and the conditions required for fusion to occur (extremely high temperature and pressure) are difficult to attain. That means that any deviation away from these conditions will result in a rapid ending of the reaction. There are no circumstances in which the plasma fusion reaction can 'run away' or proceed into an uncontrollable or critical condition.

As with conventional nuclear (fission) power, fusion power stations will produce no 'greenhouse' gases and will not contribute to global warming.

As with fission, fusion is a nuclear process; the fusion power plant structure will become radioactive, because of the action of the energetic fusion neutrons on material surfaces. However, this activation decays rapidly and the time span before it can be reused and handled can be minimised (to around 50 years) by careful selection of low-activation materials. In addition, unlike fission, there is no radioactive 'waste' product from the fusion reaction itself. The fusion by-product is helium – an inert and harmless gas.

A defining moment in the development of fusion power came in the 1950s when, flushed with the apparent success of early experiments, a respected researcher made the prediction that fusion power would be developed within 50 years. This statement has come back to haunt the programme 50 years later, as the energy source is still at least 50 years away from commercialisation. Public incredulity is understandable, but the delay is a direct result of the complex and specialised challenges to be faced in developing this potentially attractive and very long-term energy source. However, there is steady progress towards the goal.

10.3 Fusion for power

To harness fusion for power, more efficient fusion reactions than those at work in the sun are chosen. The reactions chosen are those between the two heavy forms of hydrogen: deuterium (D – containing one proton and one neutron in its core) and tritium (T – containing one proton and two neutrons). The common form of hydrogen (sometimes called protium) has no neutrons. If forced together, the deuterium and tritium nuclei fuse and then break apart to form a helium nucleus (two protons and two neutrons) and an uncharged neutron. The excess energy from the fusion reaction (released because the products of the reaction are bound together in a more stable way than the reactants) is mostly contained in the free neutron.

10.3.1 Conditions for the reaction

On earth, fusion occurs at a sufficient rate only at very high energies (temperatures). Temperatures greater than 100 million Kelvin are required (one Kelvin is similar to 1 °C, but Kelvins start from a temperature known as 'absolute zero' which is −273 °C. The freezing point of water, 0 °C is therefore 273 K).

At these extreme temperatures, a gas comprising mixed deuterium and tritium (D–T gas) becomes a plasma (a hot, electrically charged gas). In a plasma, the atoms become separated and electrons have been stripped from the atomic nuclei. For the positively charged ions to fuse, their temperature (or energy) must be sufficient to overcome their natural charge repulsion. In order to harness fusion energy, scientists and engineers are learning how to control very-high-temperature plasmas.

Plasma is an intriguing state of matter. The development of nuclear fusion has given rise to a new branch of physics examining the properties and uses of such plasmas. It is now known that matter exists in the 'plasma' state on earth in some circumstances, such as a flame or lightning. Outside the earth they are in fact the most common form of matter – plasmas comprise more than 99 per cent of the visible universe and permeate the solar system, interstellar and intergalactic environments.

Lower-temperature plasmas are now widely used, and occur for example in fluorescent tube lighting, and are used in industry, for example, in semi-conductor manufacture or spraying thin-film coatings.

However, the plasma in a fusion reactor would be a very different issue and the control of high-temperature fusion plasmas presents several major science and engineering challenges: how can a plasma be heated to more than 100 million Kelvin and how can such a plasma be confined, and maintained, so that fusion reaction can be established? It is clear that heating and confining the plasma will initially require a large amount of energy to be input to the process – an input that should be far outweighed when the fusion process begins (and passes 'breakeven'), but one that presents a challenge for developers.

Being formed by charged particles (ions and electrons), a plasma is affected by long-range electric and magnetic forces. As a consequence, plasma – and specifically magnetically confined plasma – can host an extremely rich mix of oscillations and plasma waves, covering sound, electrostatic, magnetic and electromagnetic waves. Depending on local plasma parameters, plasma waves can propagate, get dumped (absorbed), be reflected or even be converted to different plasma waves.

In general, plasma waves carry energy, so that wave absorption involves energy transfer. Their energy is then in most cases converted to the higher temperature of the absorbing medium. Wave absorption is extremely efficient if the wave frequency is resonant with some of the fundamental oscillations of the medium. However, significant heating can occur even at non-resonant frequencies – witness the widespread everyday use in microwave ovens where magnetron devices produce electromagnetic waves which heat by cyclically turning over the water molecules in food, rather than resonating with them.

10.4 Fusion parameters

In the early days a number of different schemes were investigated for confining the hot, charged plasma and various configurations tests. Initial investigations focused on linear devices, but loss of particles from the ends of these machines quickly led to experiments in which the field was 'wrapped round' to form a torus.

During the first 25 years or so, a considerable number of different experiments were built worldwide, to try out the different confinement schemes and to try to understand the underlying plasma physics. Many issues had to be understood.

During this early generation of experiments it became clear that larger and larger plasmas were needed. Elimination of the weaker or more far-fetched options left

- the tokamak, which shows the best results in a machine of given size and has the simplest magnetic confinement scheme, but is basically pulsed and prone to plasma disruptions,
- the stellarator, which inherently operates in steady state and is disruption free but uses a complex magnet and plasma geometry, and
- inertial fusion, based on firing high-energy particles or light at frozen fuel micropellets.

For the last approach, making a useful reactor is now seen to depend on focusing the energy input to the plasma in time and space, and in raising the electrical efficiency of the energy/particle beam systems providing that energy so the overall plant can produce an adequate net electrical power output. In addition, this approach faces the same overall challenges as magnetic confinement fusion, namely safe, economically competitive operation with low environmental impact.

In a reactor, the plasma would be held in a magnetic confinement device often known as a tokamak, which is a transliteration of the Russian word Токамак that itself comes from the Russian letters for the words 'toroidal chamber in magnetic coils'. This is a device that produces a toroidal (doughnut-shaped) magnetic field. It was invented in the 1950s by Igor Yevgenyevich Tamm and Andrei Sakharov.

Tokamaks utilise an ingenious scheme that addresses the two challenges of confining and heating the plasma. A huge electric current is induced in the plasma to heat it and to complement the confining magnetic field. The electric current produces heat thanks to the 'Joule Effect', a phenomenon familiar to us in everyday items such as electric ovens, irons or light bulbs. In these household appliances the electric current usually does not exceed a few amperes. Electric currents can also produce strong magnetic fields, an effect which is used in, for example, magnetic cranes, and in fact in all electric motors. Hundreds or even thousands of amperes of electric current can flow in industrial electromagnets. However, a large tokamak may induce millions of amperes into a plasma in order to heat and confine it.

The first tokamaks in the early 1960s held the plasma in a toroidal shape with a large hole in the centre, rather like a car tyre. Tokamaks like JET in the UK, with fatter-shaped plasmas, have followed.

Three parameters (plasma temperature, density and confinement time) need to be simultaneously achieved for sustained fusion to occur. The product of these is called the fusion (or triple) product and, for D–T fusion to occur, this product has to exceed a certain quantity, which is derived from the so-called Lawson Criterion, named after British scientist John Lawson who formulated it in 1955. Attaining conditions to satisfy the Lawson criterion ensures the plasma exceeds 'breakeven' – the point where the fusion power that is given out exceeds the power required to heat and sustain the plasma.

For sustained fusion to occur, the following plasma conditions need to be maintained simultaneously.

- The plasma temperature (T) must be raised to 100–200 million Kelvin for the deuterium–tritium reaction to occur. Other fusion reactions that may be sought (e.g. D–D, D–He_3) require even higher temperatures.
- The energy confinement time (t) must be at least 1–2 s. The energy confinement time is a measure of how long the energy in the plasma is retained before being lost. It is officially defined as the ratio of the thermal energy contained in the plasma and the power input required to maintain these conditions. The confinement time increases dramatically with plasma size, because large volumes retain heat much better than small volumes. The ultimate example is the sun, whose energy confinement time is massive.
- The density of fuel ions (the number per cubic metre) must be sufficiently large for fusion reactions to take place at the required rate. The fusion power generated is reduced if the fuel is diluted by impurity atoms or by the accumulation of helium ions from the fusion reaction itself. As fuel ions are burnt in the fusion process they must be replaced by new fuel and the helium product (sometimes known as the 'ash') must be removed. The necessary density of fuel ions in the plasma (n) must be 2–3 × 1020 particles per cubic metre (approx. 1/1000 g/m^3).

In practice the confinement and reaction are somewhat more complicated. Magnetic 'confinement' is in practice a force balance – a gradient in 'gas' pressure in the plasma is supported by the force resulting from the interaction of the local magnetic field with a local element of the plasma current, and the pressure is highest in the plasma interior (pressure is constant on each nested flux surface). However, the confinement is not absolute. In reality, collisions between particles cause them to drift (or diffuse) out of the magnetic confinement.

Furthermore the plasma is prone to instabilities. These are linked to the number of times the field lines on a given flux surface rotate around the minor cross section of the torus, compared to the number of times they traverse the major cross section. These 'magnetohydrodynamic (MHD) instabilities' can be controlled by careful design, but they have a tendency to be often present to some degree, triggered by, for instance, imperfections in coil manufacture and alignment. The correction coils in ITER are an example of measures taken against such instabilities.

The plasma is also prone to instabilities driven by trapped particles. Although most particles spiral round the field lines around the major circumference of the machine, some are travelling in such particular directions and with particular velocities that they are reflected at some point and reverse their flow direction. If this happens once in the torus it usually happens again at another point on the orbit of the particle, causing such particles to oscillate backwards and forwards around the torus, and providing a population of particles which can distort plasma behaviour.

Electromagnetic waves in plasma usually transfer energy to particles by collisionless damping. However if the population of particles has a strong component with higher velocities than others, and the wave velocity is below that, the wave can be amplified by the plasma causing a thermal instability, which drives energy out of the plasma. Such population distortions are caused by the heating systems themselves.

Making the most of the tokamak therefore revolves around minimising the instabilities and losses while maximising the ability of the heating to raise plasma temperature.

Losses can be reduced, and the ability of the plasma to trap heat can be improved significantly, by making the plasma larger (so that conduction losses are reduced).

10.5 Scientific feasibility

The 1970s and 1980s saw the start of operation of the most advanced magnetic confinement devices yet built, and they are at or nearing the end of their lives today. They have given a huge number of data pointing the way to optimising the plasma physics of the next step but have tantalisingly left unanswered whether large improvements in confinement are still possible on that next step, or on the contrary whether they will struggle to work well even at their nominal operating points. The ITER device must therefore be designed to operate and satisfy its objectives anywhere within that range.

10.5.1 The UK contribution

The UK contributes to fusion research in two ways: through the UK's own programme and through our contribution to the Joint European Torus (JET) project which is Euratom's flagship experiment.

JET is the Joint European Torus, sited at Culham in the UK. Work began at JET in 1983. On 1 January 2000, the European Fusion Development Agreement (EFDA) took over the operation of the JET machine after more than seventeen years of successful research.

10.5.2 Joint European Torus

JET is a tokamak and is the largest nuclear fusion experimental reactor yet built. Construction was started in 1978 and the first experiments began in 1983. In November 1991, JET became the first experiment to produce controlled fusion power. Further experiments carried out since then have provided useful information about the parameters needed for ITER.

JET is equipped with remote handling facilities to cope with the radioactivity produced by deuterium–tritium fuel, which is the fuel proposed for the first generation of fusion power plants. Pending construction of ITER, JET remains the only large fusion reactor able to use this fuel mix. In 1997 JET operations included successful experiments using the mixed deuterium–tritium fuel, and reached a record 16 MW of fusion power. The same experiment achieved a value of $Q = \sim 0.7$ where Q is the ratio of fusion power to input power. A self-sustaining nuclear fusion reaction would need a $Q > 1$, which was achieved at Japan's JT-60 tokamak the following year in 1998.

In December 1999 the JET Joint Undertaking came to an end. The United Kingdom Atomic Energy Authority (UKAEA) took over the safety and operation of the JET facilities on behalf of its European partners. The experimental programme is, as of 2000, being coordinated by the European Fusion Development Agreement (EFDA) Close Support Unit.

JET operated throughout 2003, culminating in experiments using small amounts of tritium. For most of 2004 it was shut down but the reactor was restarted in 2005 with upgraded capabilities, and in October 2006 it reached a new level of heating power at 30 MW. In the future it is possible that JET-EP (Enhanced Performance) will further increase the record for fusion power.

10.5.2.1 Heating the plasma

One of the main requirements for fusion is to heat the plasma particles to very high temperatures or energies. The following methods are typically used to heat the plasma – all of them are employed on JET.

Ohmic heating and current drive

Currents up to 5 million amperes (5 MA) are induced in the JET plasma – typically via the transformer or solenoid. As well as providing a natural pinching of the plasma column away from the walls, the current inherently heats the plasma by energising plasma electrons and ions in a particular toroidal direction. A few megawatts of heating power is provided in this way. However, at millions of degrees and above, plasma conducts electricity far too well, with very little resistance – which also means that not enough heat is produced by the Joule Effect. The unit of electric resistance is the Ohm, so plasma physicists usually say 'Ohmic heating is ineffective at high temperatures' where the word 'high' refers to the hundreds of millions of degrees required for burning plasmas. In order to attain the target temperatures some sort of 'additional heating' is required to supplement the 'Ohmic heating' (as a matter of fact, eventually the 'additional heating' plays a dominant role). Neutral particle beams and resonant electromagnetic waves can do this job.

Neutral beam heating

Beams of high-energy neutral deuterium or tritium atoms are injected into the plasma, transferring their energy to the plasma via collisions with the plasma ions. The neutral beams are produced in two distinct phases. First, a beam of energetic ions is produced by applying an accelerating voltage of up to 140,000 V. However, a beam of charged ions will not be able to penetrate the confining magnetic field in the tokamak. Thus, the second stage ensures the accelerated beams are neutralised (i.e. the ions turned into neutral atoms) before injection into the plasma. In JET, up to 21 MW of additional power is available from the NBI heating systems.

Radio-frequency heating

As the plasma ions and electrons are confined to rotating around the magnetic field lines in the tokamak, electromagnetic waves of a frequency matching the ions or electrons are able to resonate wave power into the plasma particles. As energy is transferred to the plasma at the precise location where the radio waves resonate with the ion/electron rotation, such wave heating schemes have the advantage of being localised at a particular location in the plasma. In JET, eight antennae in the vacuum vessel propagate waves in the frequency range of 25–55 MHz into the core of the plasma. These waves are tuned to resonate with particular ions in the plasma, heating them up. This method can inject up to 20 MW of heating power.

Waves can also be used to drive current in the plasma by providing a 'push' to electrons travelling in one particular direction.

Self heating of plasma

The helium ions (alpha particles) produced when deuterium and tritium fuse remain within the plasma's magnetic trap for a time until they are pumped away. The neutrons (being neutral) escape the magnetic field and their capture in a future fusion powerplant will be the source of fusion power to produce electricity. The fusion energy contained within the helium ions heats the D and T fuel ions (by collisions) to keep the fusion reaction going. When this self-heating mechanism is sufficient to maintain the required plasma temperature for fusion, the reaction becomes self-sustaining (i.e. no external plasma heating is required). This condition is referred to as ignition.

10.5.2.2 Plasma confinement

Since a plasma is made of charged particles (positive ions and negative electrons), powerful magnetic fields can be used to isolate the plasma from the walls of the containment vessel, enabling the plasma to be heated to temperatures in excess of 100 million Kelvin. This isolation of the plasma reduces the conductive heat loss through the vessel and also minimises the release of impurities from the vessel walls into the plasma that would contaminate and further cool the plasma by radiation.

In a magnetic field the charged plasma particles are forced to spiral along the magnetic field lines. The basic components of the tokamak's magnetic confinement system are:

- The toroidal field. This produces a field around the torus. It is maintained by magnetic field coils surrounding the vacuum vessel. The toroidal field provides the primary mechanism of confinement of the plasma particles.
- The poloidal field. This produces a field around the plasma cross section. It pinches the plasma away from the walls and maintains the plasma's shape and stability. The poloidal field is induced both internally, by the current driven in the plasma (one of the plasma heating mechanisms), and externally, by coils that are positioned around the perimeter of the vessel.

The main plasma current is induced in the plasma by the action of a large transformer. A changing current in the primary winding or solenoid (a multi-turn coil wound onto a large iron core in JET) induces a powerful current (up to 5 million amperes on JET) in the plasma, which acts as the transformer secondary circuit.

Other, non-magnetic plasma confinement systems are being investigated – notably inertial confinement or laser-induced fusion systems.

The most important spatial characteristics of a tokamak plasma are its 'profiles' which show how physical quantities change along the plasma radius, from the plasma centre to the plasma edge. Among the properties studied and measured are the plasma temperature profile, plasma density profile, magnetic field profile etc. These days, neutral beams as well as electromagnetic waves are used to control and modify the plasma profiles by a proper targeting of the additional energy deposition.

This technique is sometimes referred to as 'plasma tailoring' and proves extremely efficient in achieving better plasma performance. The technique can also create completely new regime conditions, for example by generating a so-called Internal Transport Barrier which provides improved plasma confinement.

The heating and current drive facilities have an even greater mission when applied as actuators (acting powers) in the JET Real Time Control. Powerful actuators can be used to automatically counteract plasma instabilities or to safeguard an intended change in plasma parameters. In this respect, additional heating and current drive will almost certainly be used in future reactors with burning plasmas.

10.5.2.3 Measuring the plasma

Measuring the key plasma properties is one of the most challenging aspects of fusion research.

Knowledge of the important plasma parameters (temperature, density, radiation losses etc.) is very important in increasing the understanding of plasma behaviour and designing, with confidence, future devices. However, as the plasma is contained in a vacuum vessel and its properties are extreme (extremely low density and extremely high temperature), conventional methods of measurement are not appropriate.

Consider how you would measure, for example, inner plasma temperature. One cannot simply put a sensitive element inside the hot plasma – not only would it evaporate, but the experiment would be lost as the plasma would cool down and become impure. Thus, plasma diagnostics have to be very innovative and often measure a physical process from which information on a particular parameter can be deduced.

First, the plasma can be observed from the outside, applying as many different methods as possible and exploiting a great variety of physical phenomena, ranging from atomic effects and nuclear reactions to radiation propagation and electromagnetism. Computing methods such as tomography, which is well known for its medical applications, provide information about plasma internal properties purely from external measurements. All wavelengths of radiated waves (visible, UV waves, x-rays etc.) are also measured – often from many locations in the plasma. Then a detailed knowledge of the process which created the waves can enable a key plasma parameter to be deduced.

Second, one can send a tiny harmless probe into the plasma, like a beam of atoms, laser light or a microwave frequency, and observe its behaviour in the hot plasma. In both cases, a good understanding of the physics underlying the measurements is essential to get sensible results.

Measurement techniques can be categorised as active or passive. In active plasma diagnostics, the plasma is probed (via laser beams, microwaves, probes etc.) to see how the plasma responds. For instance, in interferometers, the passage of a microwave beam through the plasma will be slowed by the presence of the plasma (compared to the passage through vacuum). This measures the refractive index of the plasma from which the density of plasma ions/electrons can be interpreted. With all active diagnostics, it must be ensured that the probing mechanism does not significantly affect the behaviour of the plasma.

A single vertical cross section of the plasma is sufficient for knowing more about the state of the whole plasma volume as the cross section does not vary significantly around the tokamak, in its toroidal direction. As a matter of fact, any local disturbance is immediately spread along the magnetic field lines – plasma particles move freely in this direction. Consequently, only very few fast diagnostic systems (e.g. magnetic diagnostics) monitor the toroidal irregularities. On the contrary, it is essential to measure the plasma's vertical cross section in as much detail as possible, to determine plasma profiles in the direction perpendicular to the toroidal magnetic field. The limiting factor in this is the number and position of available ports (windows into the plasma). Due to this limitation, a number of diagnostics have very similar geometrical set-ups, for example, the JET gamma-ray profile monitor, the soft x-ray diagnostics and the JET main bolometer system that measures total plasma radiated power.

10.5.3 Small tight-aspect ratio tokamak

The first tokamaks in the early 1960s had a fairly high 'aspect ratio' – the ratio of the major to minor radii of the plasma torus, that is, how close or far the overall shape is from a sphere.

Lower aspect ratio designs are closer to a sphere. At Culham, JET has an aspect ratio of around 2.5 which was more spherical than other designs of the time and demonstrated the advantages of elongated and 'D' shaped plasmas.

The START (Small Tight-Aspect Ratio Tokamak) experiment at the Culham Science Centre began operation in 1991 and was designed to take this trend in plasma shape to its limits. It employs a cylindrical vacuum vessel 2 m diameter and 2 m high, and plasmas are obtained at aspect ratios as low as 1.25. START produced an even more compact plasma, like a cored apple.

Numerical equilibrium studies had shown that spherical tokamak plasmas are naturally elongated and D-shaped and moreover have a natural system for power and particle exhaust, an advantage for an eventual power plant. Significantly, it is found that the toroidal magnetic field (supplied by the centre column) required for plasma stability is a factor of 10 less for the ST than for the conventional case – implying a substantial gain in efficiency. Further advantages are implied by theories of particle behaviour at low aspect ratio, suggesting improved plasma heating and other benefits.

10.5.3.1 Experimental results from START

Built initially as a low-cost experiment to test the theoretical predictions made for tight-aspect ratio tokamak systems, START quickly became the world's leading spherical tokamak experiment. Many fusion scientists from countries all over the world, including Brazil, Japan, USA and Russia, have visited Culham to take part in research on START. The high temperatures achieved (more than 10 million Kelvin) in combination with its small size and low construction cost created worldwide interest in this novel approach, and many similar devices are now being built around the world.

START has provided the world's first experimental results on hot (Te $\sim$ 500 eV or 5 million °C) spherical tokamak plasmas.

START began operation in January 1991 and immediately gave promising results, with measurements of high electron temperatures. High plasma densities can be obtained – the START 'operating space' is at least as large as in conventional tokamaks, and energy confinement in START plasmas is better than predicted by most confinement scalings. Since 1991 considerable development and improvement to both device and diagnostics have taken place, with a significant increase in plasma size. Plasma current has increased from 20,000 to over 300,000 A. An unexpected but important advantage of this is that the current-terminating 'major disruptions' normally observed in conventional tokamaks appear to be largely self-stabilising on START, and although terminations can occur they are more easily avoided.

Spherical tokamaks are theoretically predicted to have higher plasma containment efficiency than conventional tokamaks. A measure of this is ß, the ratio of plasma pressure to magnetic field pressure. To test this in START, a Neutral Beam Injector (NBI) was loaned to Culham by the Oak Ridge National Laboratory, USA, and used to provide additional heating of the START plasma. A new world record in ß was achieved, reaching 40 per cent on START, trebling the previous record of 12.6 per cent on the DIII-D tokamak at General Atomics, San Diego.

In 1997 and 1998 'improved confinement' modes of operation (similar to the H-mode observed in conventional tokamaks) were achieved. The clearest evidence of this is given by the high-speed video images that show a very sharply defined edge to the plasma, indicating a low level of turbulence and hence good confinement. These exciting results are still being analysed.

The START project has shown that the spherical tokamak appears to behave at least as well as the conventional tokamak. It has similar operating regimes, confinement and stability but is resilient to the current-terminating major disruption and has a potentially useful 'natural' exhaust system. Coupled with its simplicity and the substantial increase in efficiency, demonstrated by the trebling of the tokamak world record value of ß achieved by START, there is every reason to continue research in this area. Inspired by START results, next-stage devices are being operated.

10.5.4 Mega amp spherical tokamak

At Culham, a new tokamak called MAST (Mega Amp Spherical Tokamak) has taken over from START, which completed its experimental programme in March 1998.

The new machine took two years to design and a further two years to construct. It has been fully commissioned and the experimental programme commenced in December 1999.

The first spherical tokamak plasmas obtained by merging and compression were produced on the new MAST magnetic fusion facility at Culham Science Centre in the last few days of the millennium. This was an extremely encouraging start to the programme of work on the world's largest spherical tokamak, with the production of plasmas over 3 m in diameter.

The installation of the centre column and engineering commissioning went very smoothly, well ahead of schedule. Initiation and production of toroidal plasmas for a wide range of aspect ratios (ratio of major to minor radius), with currents of up to 1/3 of a million amperes, was found to be very easy and the first additional heating experiments have begun. A large number of diagnostic systems allows deduction of the nature of these spherical plasmas and how they scale from the much smaller, but very successful, START experiment.

10.5.5 The way ahead

The success of JET, in terms of optimising plasma stability and confinement, has led to the design of the next-step device: the International Thermonuclear Experimental Reactor (ITER). ITER is an international collaboration with seven partners (EU, Japan, USA, South Korea, Russia, China and India) and is a more advanced, larger version of JET. It will be capable of producing 500 MW of fusion power (ten times that needed to heat the plasma). In comparison, JET can only produce fusion power that is ~70 per cent of the power needed to heat the plasma.

After much political debate, the go-ahead to build ITER at Cadarache in France was given in June 2005 (see Panel 10.1). ITER will take ten years to build and should operate from 2015.

The so-called fast track to commercial fusion power is a strategy designed to ensure that a demonstration fusion power station puts electricity into the grid in 30 years time. During the operation of ITER, a parallel materials testing programme will be undertaken – developing and assessing the materials needed for a power plant. The experience from both these facilities will enable the first demonstration power plant to be operational, in around 30 years.

ITER will have the same magnetic geometry as JET. Similar but much bigger than JET and with the addition of a number of key technologies essential for a future power station, ITER will be able to operate for very much longer periods (pulses lasting over 500 s) and will help to demonstrate the scientific and technological feasibility of fusion power. Most importantly, it will be the first fusion device designed to achieve sustained burn – at which point the reactor becomes self-heating and productive.

During its operating life, ITER must not only address challenges in producing and managing the plasma and passing the breakeven point for the fusion reaction. It must also be the qualifying ground for many of the techniques and technologies that will be required for fission reactors that follow it, specifically a reactor known as 'Demo', likely in around 30 years, that will be a 'demonstration' reactor for potential deployment (e.g. ITER will test and confirm design solutions for the 'blankets' required to produce tritium from lithium). ITER provides the only good opportunity to identify the best engineering solutions to generate sufficient tritium fuel in a realistic fusion reactor environment with minimum contamination of the high-grade coolant needed for steam generation. However, due to ITER's relatively low power level, the testing

Panel 10.1 Progress to the ITER

1988 The Conceptual Design Activity for the International Thermonuclear Experimental Reactor (ITER), the successor to TFTR, JET and JT-60, began. Participants were EURATOM, Japan, Soviet Union and United States. It ended in 1990.

1992 The Engineering Design Activity for the ITER began. Participants were EURATOM, Japan, Russia and United States. It ended in 2001.

1997 Groundbreaking ceremony held in the USA for the National Ignition Facility (NIF). Combining a field-reversed pinch with an imploding magnetic cylinder resulted in the new Magnetised Target Fusion concept in the US. In this system a 'normal' lower-density plasma device was explosively squeezed using techniques developed for high-speed gun research.

1999 The United States withdrew from the ITER project.

2001 Building construction for the immense 192 beam 500 TW NIF project is completed and construction of laser beamlines and target bay diagnostics commences. The NIF is expected to take its first full system shot in 2010.

2001 Negotiations Meeting on the Joint Implementation of ITER begins. Participants were Canada, European Union, Japan and Russia.

2002 European Union proposed Cadarache in France and Vandellos in Spain as candidate sites for ITER while Japan proposed Rokkasho.

2003 The United States rejoined the ITER project, and China and Republic of Korea newly joined while Canada withdrew.

2003 Cadarache in France was selected as the European Candidate Site for ITER.

2004 The United States dropped its own project, the Fusion Ignition Research Experiment (FIRE), to focus resources on ITER.

2005 Following final negotiations between the EU and Japan, ITER chose Cadarache over Rokkasho for the site of the reactor. In concession, Japan was made the host site for a related materials research facility and was granted rights to fill 20 per cent of the project's research posts while providing 10 per cent of the funding.

will take a considerable time – about 15 years of machine operation. ITER can be run to end of life using externally supplied tritium (one source is the heavy-water coolant from today's CANDU fission reactors), but all the subsequent machines must rely on tritium generated by themselves.

In the early days of the ITER design, one plan was to make the main structure of the machine from low-activation materials. However, to construct ITER it must satisfy licensing authorities that it will behave as predicted and not release radioactive effluents. The low-activation materials under consideration are

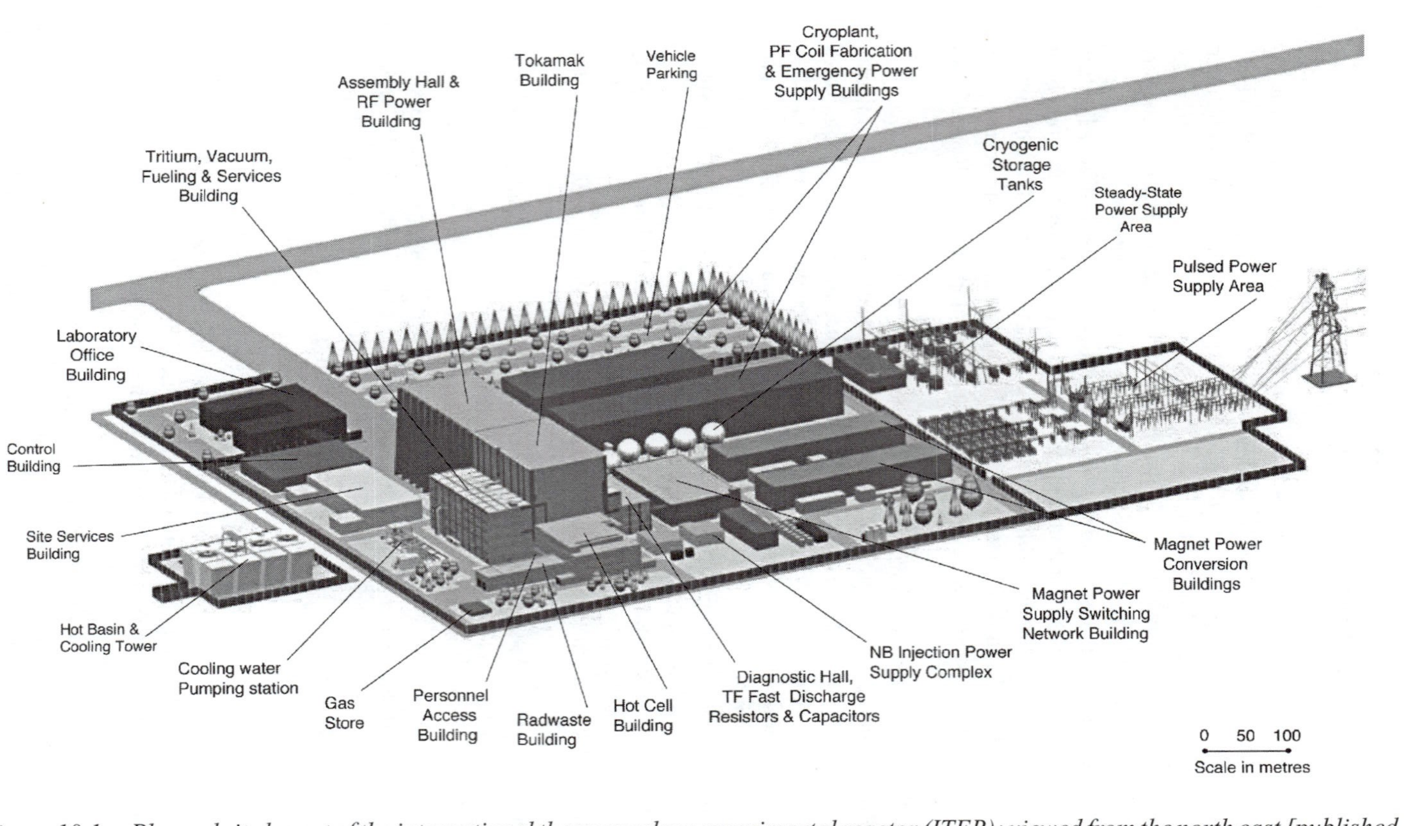

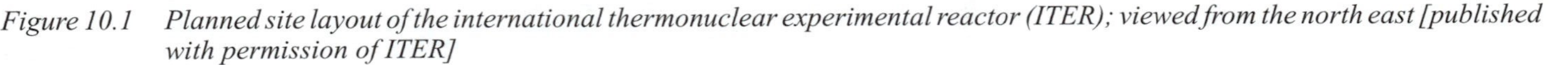
Figure 10.1 *Planned site layout of the international thermonuclear experimental reactor (ITER); viewed from the north east [published with permission of ITER]*

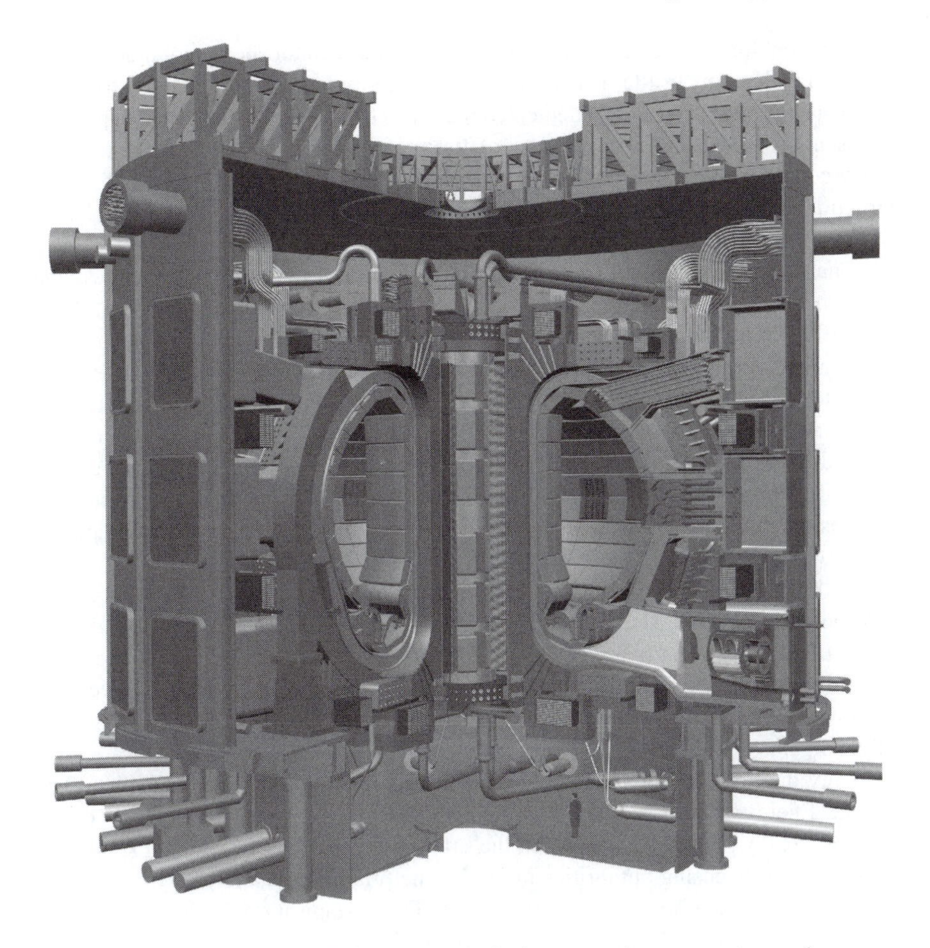

Figure 10.2 Cutaway of the international thermonuclear experimental reactor design [published with permission of ITER]

steels with a martensitic crystal form in which precursor elements which lead to long-lived radioactive isotopes are replaced by elements with a similar material function – so-called isotopically tailored steels, vanadium-based alloys and silicon carbide.

Although there is some experience of how some of these materials will behave under the type of irradiation that arises from fission neutrons, and the results look promising for the future, there are insufficient data on which to build a machine and guarantee its integrity in a nuclear environment. Only stainless steel with an austenitic crystal form, used widely in the fission programme, has accumulated sufficient data to be qualified as a structural material for ITER.

During the life of ITER, low-activation materials need to be qualified so that they can be licensed for use as structural materials on the next machine after ITER and beyond. Although austenitic steels can be proven to withstand the ITER nuclear environment for the operational lifetime of ITER, they cannot operate for much longer before they become brittle. ITER itself cannot run long enough to qualify structural materials for later devices so a dedicated materials testing facility operated in parallel is essential. Based on the experience of fission, it takes about 20 years to develop the necessary materials database and consistent manufacture so, to be ready for construction of a demonstration reactor to follow ITER, this materials test facility has to begin operation in around 2010.

The limited operational lifetime of ITER also does not allow it to address the issue of the endurance of the breeding blanket designs developed and partly tested on ITER. Thermohydraulic and neutronic behaviour, breeding capability and tritium removal techniques can all be confirmed, but the life of the component can only be predicted based on the concurrent materials development programme. Without it, the best guarantees for the operating life of DEMO in-vessel components cannot extend beyond a few months, and the consequences of a mistaken design or material choice for DEMO based on a lack of information would be extra expense and delay, as well as considerable unnecessary damage to the reputation of fusion as a reliable power source.

Following the decision on the ITER construction site at the end of June 2005, it is now necessary to finalise the Joint Implementation Agreement (JIA), have it signed and ratified by each party (since it is an international agreement under international law) and establish the ITER organisation as a legal entity. This phase is necessary before the Joint Implementation Agreement is eventually signed or ratified and the ITER organisation is set up.

Only then can an application for a licence to construct ITER be launched. The Atomic Energy Commissariat (CEA), acting for France as a potential ITER host country, has already prepared a dossier on the safety aspects of ITER, which it submitted to the French licensing authorities in 2002. The preliminary safety report on ITER will be submitted as soon as possible by the ITER organisation. It will describe the actual design to be built and the safety of the facility.

This will be followed by an estimated seven-year construction phase during which the first large hardware contracts are launched (some sub-component procurements uncritical for licensing may even be launched earlier) and, eventually, all sub-systems are assembled and commissioned.

Of the estimated 21-year operation phase, one year of integrated system commissioning will be followed by 10 years of operation aiming primarily at establishing the optimum physics of a power reactor and determining the best operating mode to obtain the most relevant tritium-breeding blanket testing. This will be followed by a 10-year operation phase to exploit those conditions.

Finally there will be a decommissioning phase, of which the first six years will be the final responsibility of the ITER project. This will encompass deactivating the plant by removing all tritiated materials, activated corrosion products and radioactive dust, as well as in-vessel components, followed by about 20 further years (minimum) under the responsibility of the host country (France and the EU) allowing for radioactive decay, and a further six-year period of dismantlement and disposal of the remaining plant.

10.6 The cost of fusion power

The cost to the electricity consumer for the ITER experimental reactor is estimated at about €0.40/y per person for 30 years for Europeans.

It is not yet possible to say whether nuclear fusion based on magnetic confinement will produce a competitive energy source. Due to completed and ongoing experiments, the signs are encouraging, much is understood about the basic science and there is a clear path to its technical realisation. Building ITER will advance this knowledge immeasurably, particularly with regard to optimisation of the physics performance parameters, technology to be used and appropriate manufacturing techniques to be adopted, and place humanity in a much better position to make an informed decision on whether it is worth going further.

The design of a fusion power reactor based on magnetic fusion has still to be worked out. However, irrespective of the design, something can be said about the economics of fusion reactors by making some basic assumptions:

- the reactors will use D+T reactions and the water or helium coolants will drive steam or gas turbines;
- fusion power plants will preferably have a similar generating capacity to today's coal and nuclear plants – this means that power delivered to the network per plant should be in the range 500–1,500 MW (electric);
- the economic rules that apply in today's environment to coal and nuclear plants will apply to fusion – the cost of raising money for the investment must be considered, and future benefits and costs must be discounted to estimate an appropriate cost (and subsequently, price) of electricity;
- lifetime costs (i.e. including decommissioning and waste disposal) must be included.

Although ITER is not a full power plant it incorporates most of the same equipment at roughly the right scale. Some additional power-generation equipment will be needed and higher heat tolerance and longer endurance nuclear components must be used, and the physical size may be a little larger, but there will be economies of scale due to series production of identical plants, so the cost of ITER is to some extent representative of a fusion power plant.

Using these assumptions, a reactor cost model has been developed around the following equations:

$$P_{gen} = P_{fus}(0.2 + 0.8M)h_{th} - P_{internal} - P_{fus}/Q/h_{hcd},$$

where P_{gen} is the power delivered to the network; P_{fus} is the fusion power; M is the blanket neutron power amplification; h_{th} is the thermal conversion efficiency; $P_{internal}$ is the in-house electrical power consumption (except heating and current drive); Q is the plasma power amplification (of heating and current drive power); h_{hcd} is the efficiency of conversion of electricity to heating and current drive power to the plasma.

$$\text{Cost of electricity (COE)} = \frac{\text{Discounted project cost}}{\text{Discounted electricity to the network}}$$

$$\text{Discounted project cost} = \text{Direct} + \text{Indirect} + \text{Construction Interest} + \text{Maintenance}$$

Direct costs are taken as some factor times ITER costs.

Indirect costs, interest during construction and maintenance (including decommissioning) are calculated as a factor of direct costs.

$$\text{Discounted electricity to the network} = \text{Sum over operating life}\left[\frac{P_{gen}A_n}{(1+d)(n-0.5)}\right],$$

where A_n is plant availability in year n; d is the discount rate.

Domestic electricity price = 2.5 times the cost of electricity production.

Industrial electricity price = 0.462 times domestic prices (these values are currently appropriate in Europe).

The results show that, provided a power reactor cost can be maintained near the cost of ITER, there are good chances that the price of electricity produced in future by fusion will be comparable to that achieved today from coal and fission, especially when environmental costs not currently factored into those energy sources are included.

The model also shows the importance of the various parameters (both technical and economic) to achieving that goal.

10.7 Following ITER

ITER is planned to operate at a nominal fusion power of 500 MWt. If DEMO (the next device after ITER, and the first to generate electricity) is to be a device of approximately similar physical size (and hence cost), its fusion power level has to be increased by about a factor of four, so that the electrical power potentially delivered to the network will be in the range of 500 MWe, typical of one of today's power stations (albeit rather a small one). The general level of heat fluxes through the walls will be about four times higher than in ITER, and plasma performance needs to be improved to gain this four-fold increase.

Calculations show that this performance could be achieved with an increase of around 15 per cent in ITER linear dimensions, and a 30 per cent increase in the plasma density above the nominal expected to be confined by the basic magnetic fields on ITER. It then remains for enough to be learnt on the ITER blanket test beds to allow the DEMO blanket to be designed to withstand four times the ITER steady heat loads on those components.

If these systems work successfully on DEMO, DEMO itself can be used as a prototype commercial reactor creating a 'fast track' to fusion. This would accelerate the availability of fusion as an energy option by about 20 years. A further step would no doubt subsequently be made for the first-of-a-series commercial-sized fusion power reactor (PROTO), doubling the electrical power by increasing linear machine dimensions by less than 10 per cent, without assuming any improvement in physical behaviour.

Assuming plant capital cost scales with the tokamak volume, one can expect DEMO capital costs in the region of €14/We based on the cost estimates for ITER. Those of PROTO will then be typically €8/We and with subsequent economies of series production of fusion plants, capital costs could reduce to €4/We. This should be compared to today's fission and coal plants at €3/We and €1.5/We, respectively.

However, the capital costs of today's coal plants do not include costs to mitigate environmental damage, nor do any of the above costs include the fuel, operating and decommissioning costs, which for coal are typically comparable to the capital costs and should be lowest for fusion.

Panel 10.2 Tokamak reactors

TFTR

The Tokamak Fusion Test Reactor (TFTR) operated at the Princeton Plasma Physics Laboratory (PPPL) from 1982 to 1997. TFTR set a number of world records, including a plasma temperature of 510 million degrees centigrade – the highest ever produced in a laboratory and well beyond the 100 million degrees required for commercial fusion. In addition to meeting its physics objectives, TFTR achieved all of its hardware design goals, thus making substantial contributions in many areas of fusion technology development.

In December, 1993, TFTR became the world's first magnetic fusion device to perform extensive experiments with plasmas composed of 50/50 deuterium/tritium – the fuel mix required for practical fusion power production. Consequently, in 1994, TFTR produced a world-record 10.7 million watts of controlled fusion power, enough to meet the needs of more than 3,000 homes. These experiments also emphasised studies of the behaviour of alpha particles produced in the deuterium–tritium reactions. The extent to which the alpha particles pass their energy to the plasma is critical to the eventual attainment of sustained fusion.

In 1995, TFTR scientists explored a new fundamental mode of plasma confinement – enhanced reversed shear. This new technique involves a magnetic-field configuration that substantially reduces plasma turbulence.

JT-60

Work began at Japan's JT-60 tokamak in 1985. The torus is sited at the Naka Fusion Institute, part of the Japan Atomic Energy Agency. It is still in operation: research objectives for 2005–6 and the reactor are now being modified to carry out additional work on high-beta steady-state research. The modification of JT-60 is regarded as 'National Centralized Tokamak (NCT) Facility Program' and detailed design work is ongoing in collaboration with universities, institutes and industries in Japan.

Other tokamaks used for research and development include:

Compact Ignition Tokamak
Alcator C-Mod (MIT)
D-IIID (General Atomics)
HBT-EP (Columbia University)
ASDEX-U (Axisymmetric Diverter Experiment) (Garching, Germany)
FT-U (Italy)
Textor (Germany)
Tore Supra (France)
JRT2-M (Japan)
Triam-1M (Japan)
K-Star (Korea)
ATF (Advanced Toroidal Facility) (US)

Glossary

Actinide: An element with atomic number of 89 (actinium) or above.

Activation product: A radioactive isotope of an element (e.g. in the steel of a reactor core) created by neutron bombardment.

Activity: The number of disintegrations per unit time inside a radioactive source. Expressed in becquerels.

ALARA: As Low As Reasonably Achievable, economic and social factors being taken into account. This is the optimisation principle of radiation protection.

Alpha particle: A positively charged particle from the nucleus of an atom, emitted during radioactive decay. Alpha particles are helium nuclei, with 2 protons and 2 neutrons.

Atom: A particle of matter which cannot be broken up by chemical means. Atoms have a nucleus consisting of positively charged protons and uncharged neutrons of the same mass. The positive charges on the protons are balanced by a number of negatively charged electrons in motion around the nucleus.

Background radiation: The naturally occurring ionising radiation which every person is exposed to, arising from the earth's crust (including radon) and from cosmic radiation.

Barn: see Cross section.

Base load: That part of electricity demand which is continuous, and does not vary over a 24-h period. Approximately equivalent to the minimum daily load.

Becquerel: The SI unit of intrinsic radioactivity in a material. One becquerel measures one disintegration per second and is thus the activity of a quantity of radioactive material which averages one decay per second. (In practice, GBq or TBq are the common units.)

Beta particle: A particle emitted from an atom during radioactive decay. Beta particles may be either electrons (with negative charge) or positrons.

Biological shield: A mass of absorbing material (e.g. thick concrete walls) placed around a reactor or radioactive material to reduce the radiation (especially neutrons and gamma rays, respectively) to a level safe for humans.

Boiling-water reactor (BWR): A common type of light-water reactor (LWR), where water is allowed to boil in the core thus generating steam directly in the reactor vessel.

Breed: To form fissile nuclei, usually as a result of neutron capture, possibly followed by radioactive decay.

Breeder reactor: See Fast Breeder Reactor and Fast Neutron Reactor.

Burnable poison: A neutron absorber included in the fuel, which progressively disappears and compensates for the loss of reactivity as the fuel is consumed. Gadolinium is commonly used.

Burn: To cause fission.

Burnup: Measure of thermal energy released by nuclear fuel relative to its mass, typically Gigawatt days per tonne (GWd/tU).

Calandria: (in a CANDU reactor) A cylindrical reactor vessel which contains the heavy-water moderator. It is penetrated from end to end by hundreds of calandria tubes which accommodate the pressure tubes containing the fuel and coolant.

CANDU: Canadian deuterium uranium reactor, moderated and (usually) cooled by heavy water.

Chain reaction: A reaction that stimulates its own repetition, in particular where the neutrons originating from nuclear fission cause an ongoing series of fission reactions.

Cladding: The metal tubes containing oxide fuel pellets in a reactor core.

Concentrate: See uranium oxide concentrate (U_3O_8).

Containment: Methods or physical structures designed to prevent the dispersion of radioactive substances.

Control rods: Devices to absorb neutrons so that the chain reaction in a reactor core may be slowed or stopped by inserting them further, or accelerated by withdrawing them.

Conversion: Chemical process turning U_3O_8 into UF_6 preparatory to enrichment.

Coolant: The liquid or gas used to transfer heat from the reactor core to the steam generators or directly to the turbines.

Core: The central part of a nuclear reactor containing the fuel elements and any moderator.

Corium: Product which would result from the melting of the core components and their interaction with the structures they would meet.

Critical mass: The smallest mass of fissile material that will support a self-sustaining chain reaction under specified conditions.

Criticality: Condition of being able to sustain a nuclear chain reaction.

Cross section: A measure of the probability of an interaction between a particle and a target nucleus, expressed in barns (1barn = 10–24 cm^2).

Curie (Ci): Unit of activity, equal to 3.7×10^{10} Bq (exactly). Superseded by the becquerel (Bq).

Decay: Disintegration of atomic nuclei resulting in the emission of alpha or beta particles (usually with gamma radiation). Also the exponential decrease in radioactivity of a material as nuclear disintegrations take place and more stable nuclei are formed.

Decommissioning: The removal of a nuclear reactor or other facility from service, and the subsequent dismantling and making the site available for unrestricted use. In some cases used to refer only to the dismantling and removal of the facility.

Decontamination: The complete or partial removal of contamination by a deliberate physical, chemical or biological process. This definition is intended to include a wide range of processes but to exclude the removal of radionuclides from within the human body, which is not considered to be decontamination.

Delayed neutrons: Neutrons released by fission products up to several seconds after fission. These enable control of the fission in a nuclear reactor.

Depleted uranium: Uranium with less than the natural proportion of 0.7 per cent ^{235}U. As a by-product of enrichment in the fuel cycle it generally has 0.25–0.30 per cent ^{235}U, the rest being ^{238}U. Can be blended with highly enriched uranium (e.g. from weapons) to make reactor fuel.

Design basis: The range of conditions and events taken explicitly into account in the design of a facility, according to established criteria, such that the facility can withstand them without exceeding authorised limits by the planned operation of safety systems.

Deuterium: 'Heavy hydrogen', a stable isotope having one proton and one neutron in the nucleus. It occurs in nature as 1 atom to 6,500 atoms of normal hydrogen, which has one proton and no neutrons.

Disintegration: Natural change in the nucleus of a radioactive isotope as particles are emitted (usually with gamma-rays), making it a different element.

Dose: The energy absorbed by tissue from ionising radiation. One grey is one joule per kilogram, but this is adjusted for the effect of different kinds of radiation, and thus the sievert is the unit of dose equivalent used in setting exposure standards.

Electron Volt (eV): A unit of energy that is convenient to use on the atomic scale, equal to the amount of work required to move an electron, with its negative charge, through an electrical potential of one volt, 1.6×10^{-19} joules. Chemical reactions at the atomic scale involve energies of a few electron volts, for example, combining two molecules of hydrogen (H_2) with a single molecule of oxygen (O_2) to make two molecules of water (H_2O) releases 5 eV. Nuclear reactions release much more energy. Fission of a uranium-235 nucleus by a neutron releases 200 MeV – 200 million electron volts. The fusion of deuterium with tritium releases 17.9 MeV, which, because the mass of the deuterium and tritium is so small, is 500 times the energy released by fission, per kilogram of fuel, and 100 million times the energy released by burning coal.

Element: A chemical substance that cannot be divided into simple substances by chemical means; atomic species with same number of protons.

Embrittlement: Gradual alteration of the structure of a metal when subjected to long-term neutron bombardment, for example in a nuclear reactor. The change in structure can make it less ductile and more vulnerable to cracking. This is an ageing issue that needs to be addressed, particularly for operators of Russian-designed VVERs, which have a weld at core level that is particularly susceptible to embrittlement. The effect can be largely reversed by a heat treatment known as annealing.

Enriched uranium: Uranium in which the proportion of ^{235}U (to ^{238}U) has been increased above the natural 0.7 per cent. Reactor-grade uranium is usually enriched to about 3.5 per cent ^{235}U, weapons-grade uranium is more than 90 per cent ^{235}U.

Enrichment: Physical process of increasing the proportion of ^{235}U to ^{238}U. See also SWU.

Fast breeder reactor (FBR): A fast neutron reactor configured to produce more fissile material than it consumes, using fertile material such as depleted uranium in a blanket around the core.

Fast neutron: Neutron released during fission, travelling at very high velocity (20,000 km/s) and having high energy (*c* 2 MeV).

Fast neutron reactor: A reactor with little or no moderator and hence utilising fast neutrons. It normally burns plutonium while producing fissile isotopes in fertile material such as depleted uranium (or thorium).

Fertile (of an isotope): Capable of becoming fissile, by capturing neutrons, possibly followed by radioactive decay; for example ^{238}U, ^{240}Pu.

Fissile (of an isotope): Capable of capturing a slow (thermal) neutron and undergoing nuclear fission, for example ^{235}U, ^{233}U, ^{239}Pu.

Fissionable (of an isotope): Capable of undergoing fission: if fissile, by slow neutrons; if fertile, by fast neutrons.

Fission: The splitting of a heavy nucleus into two, accompanied by the release of a relatively large amount of energy and usually one or more neutrons. It may be spontaneous but usually is due to a nucleus absorbing a neutron and thus becoming unstable.

Fission products: Daughter nuclei resulting either from the fission of heavy elements such as uranium, or the radioactive decay of those primary daughters. Usually highly radioactive.

Fossil fuel: A fuel based on carbon presumed to be originally from living matter, coal, oil, gas. Burnt with oxygen to yield energy.

Fuel assembly: Structured collection of fuel rods or elements, the unit of fuel in a reactor.

Fuel fabrication: Making reactor fuel assemblies, usually from sintered UO_2 pellets which are inserted into zircaloy tubes, comprising the fuel rods or elements.

Fuel pin: Fuel around 1 cm in diameter and varying between 30 cm and several metres in length, containing a number of fuel pellets in a steel or zircaloy tube. Up to 300 pins are used in a fuel assembly.

Gamma rays: High-energy electromagnetic radiation from the atomic nucleus, virtually identical to x-rays.

Giga: One billion units (e.g. gigawatt, 10^9 watts or million kilowatts).

Graphite: Crystalline carbon used in very pure form as a moderator, principally in gas-cooled reactors, but also in Soviet-designed RBMK reactors.

Greenhouse gases: Radiative gases in the earth's atmosphere that absorb long-wave heat radiation from the earth's surface and re-radiate it, thereby warming the earth. Carbon dioxide and water vapour are the most important, although not necessarily those with the greatest warming effect per volume.

Grey: The SI unit of absorbed radiation dose, one joule per kilogram of tissue.

Half-life: The period required for half of the atoms of a particular radioactive isotope to decay and become an isotope of another element.

Heavy water: Water containing an elevated concentration of molecules with deuterium ('heavy hydrogen') atoms.

Heavy-water reactor (HWR): A reactor which uses heavy water as its moderator, for example Canadian CANDU (pressurised HWR or PHWR).

High-level wastes: Extremely radioactive fission products and transuranic elements (usually other than plutonium) in spent nuclear fuel. They may be separated by reprocessing the spent fuel, or the spent fuel containing them may be regarded as high-level waste.

Highly (or High)-enriched uranium (HEU): Uranium enriched to at least 20 per cent ^{235}U. That in weapons is about 90 per cent ^{235}U.

***In situ* leaching (ISL):** The recovery by chemical leaching of minerals from porous ore bodies without physical excavation. Also known as solution mining.

Ion: An atom that is electrically charged because of loss or gain of electrons.

Ionising radiation: Radiation (including alpha particles) capable of breaking chemical bonds, thus causing ionisation of the matter through which it passes and causing damage to living tissue.

Irradiate: Subject material to ionising radiation. Irradiated reactor fuel and components have been subject to neutron irradiation and hence become radioactive themselves.

Isotope: An atomic form of an element having a particular number of neutrons. Different isotopes of an element have the same number of protons but different numbers of neutrons and hence different atomic mass, for example ^{235}U, ^{238}U. Some isotopes are unstable and decay to form isotopes of other elements.

Joule (J): A unit of energy, equal to the work required to lift a one kilogram mass a distance of 0.101 metres, or to lift a one pound mass 8.8 inches. For IFE the unit megajoules (MJ), or millions of joules, is convenient. For comparison, the detonation of one kilogram (2.2 pounds) of high explosive releases roughly 4.2 MJ, the combustion of a kilogram of coal releases just over 30 MJ of energy and the National Ignition Facility's lasers will deposit 1.8 MJ of laser energy into ICF targets. Units of kilojoules (kJ), or thousands of joules; and gigajoules (GJ), or billions of joules, are also used.

Kelvin: Measure of temperature. One Kelvin is equivalent to one degree centigrade, but whereas centigrade is measured from the freezing point of water (0 °C) Kelvins are measured from a point known as 'absolute zero'. This is −273 °C.

Kilogramme: The kilogramme is a unit of mass, equal to 2.2 pounds, or 1,000 g.

Kilowatt hour: The amount of electrical energy required to provide power at one kilowatt (1,000 watts) for one hour. A typical person, working relatively hard, can generate around 200 watts of energy. Such a person would take 5 h to generate one kilowatt hour, and at current typical electricity costs, would earn from 2 to 10 cents for their five hours of effort.

Kinetic energy: Energy associated with the motion of an object, equal to $\frac{1}{2}mv^2$ where m is the object mass and v is the object speed.

Light water: Ordinary water (H_2O) as distinct from heavy water.

Light-water reactor (LWR): A common nuclear reactor cooled and usually moderated by ordinary water.

Lithium: The third element in the periodic table, lithium has three protons and either three (^{6}Li) or four (^{7}Li) neutrons. The isotope ^{6}Li provides the source of tritium for fusion power plants.

Low-enriched uranium: Uranium enriched to less than 20 per cent ^{235}U. That in power reactors is usually 3.5–5.0 per cent ^{235}U.

Magnetic-Confinement Fusion (MFE): In magnetic-confinement fusion, magnetic fields are used to keep moving ions from touching the chamber wall, allowing sufficiently high temperatures and densities to be reached to allow fusion reactions to occur.

Megawatt (MW): A megawatt is a million watts, sufficient power to light 10,000 100-watt bulbs, or enough electricity for around 3,000 households. The watt is a unit of power, equal to one joule per second. Sometimes the unit MWe, or 'megawatt electric' is used to distinguish the electrical power produced by a plant from the larger amount of heat (MWt, or megawatt thermal) required to make that electricity, due to the inherent limitation on the efficiency of converting heat to electricity.

Metal fuels: Natural uranium metal as used in a gas-cooled reactor.

Micro: one millionth of a unit (e.g. microsievert is 10^{-6} Sv).

Migration: The movement of radionuclides in the environment as a result of natural processes. Most commonly, movement of radionuclides in association with groundwater flow.

Milling: Process by which minerals are extracted from ore, usually at the mine site.

Mixed oxide fuel (MOX): Reactor fuel which consists of both uranium and plutonium oxides, usually about 5 per cent Pu, which is the main fissile component.

Moderator: A material such as light or heavy water or graphite used in a reactor to slow down fast neutrons by collision with lighter nuclei so as to expedite further fission.

Natural uranium: Uranium with an isotopic composition as found in nature, containing 99.3 per cent ^{238}U, 0.7 per cent ^{235}U and a trace of ^{234}U. Can be used as fuel in heavy-water moderated reactors.

Neutron: An uncharged elementary particle found in the nucleus of every atom except hydrogen. Solitary mobile neutrons travelling at various speeds originate from fission reactions. Slow (thermal) neutrons can in turn readily cause fission in nuclei of 'fissile' isotopes, for example ^{235}U, ^{239}Pu, ^{233}U; and fast neutrons can cause fission in nuclei of 'fertile' isotopes such as ^{238}U, ^{239}Pu. Sometimes atomic nuclei simply capture neutrons.

Nuclear reactor: A device in which a nuclear fission chain reaction occurs under controlled conditions so that the heat yield can be harnessed or the neutron beams utilised. All commercial reactors are thermal reactors, using a moderator to slow down the neutrons.

Nuclide: Elemental matter made up of atoms with identical nuclei, therefore with the same atomic number and the same mass number (equal to the sum of the number of protons and neutrons).

Oxide fuels: Enriched or natural uranium in the form of the oxide UO_2, used in many types of reactor.

Plutonium: A transuranic element, formed in a nuclear reactor by neutron capture. It has several isotopes, some of which are fissile and some of which undergo spontaneous fission, releasing neutrons. Weapons-grade plutonium is produced in special reactors to give >90 per cent ^{239}Pu, reactor-grade plutonium contains about 30 per cent non-fissile isotopes. About one third of the energy in a light-water reactor comes from the fission of ^{239}Pu, and this is the main isotope of value recovered from reprocessing spent fuel.

Pressurised-water reactor (PWR): The most common type of light-water reactor (LWR), it uses water at very high pressure in a primary circuit and steam is formed in a secondary circuit.

Pressuriser: A primary circuit component of a pressurised-water reactor which regulates the pressure in the circuit.

Primary circuit: The core and cooling loops of a light-water reactor. The primary circuit can contain fission products so in PWRs heat from the primary circuit is transferred to a secondary circuit which includes the turbine generator.

Radiation: The emission and propagation of energy by means of electromagnetic waves or particles (*cf* ionising radiation).

Radioactivity: The spontaneous decay of an unstable atomic nucleus, giving rise to the emission of radiation.

Radionuclide: A radioactive isotope of an element.

Radiotoxicity: The adverse health effects of a radionuclide due to its radioactivity.

Radium: A radioactive decay product of uranium often found in uranium ore. It has several radioactive isotopes. Radium-226 decays to radon-222.

Radon (Rn): A heavy radioactive gas given off by rocks containing radium (or thorium). ^{222}Rn is the main isotope.

Radon daughters: Short-lived decay products of radon-222 (^{218}Po, ^{214}Pb, ^{214}Bi, ^{214}Po).

RBMK: A Russian-designed reactor using a graphite moderator and light-water coolant. Short for Reaktor Bolshoi Moshchnosty Kanalny or 'high-power channel reactor'.

Reactor pressure vessel: The main steel vessel containing the reactor fuel, moderator and coolant under pressure.

Repository: A permanent disposal place for radioactive wastes.

Reprocessing: Chemical treatment of spent reactor fuel to separate uranium and plutonium from the small quantity of fission product waste products and transuranic elements, leaving a much reduced quantity of high-level waste (see Waste, HLW).

Residual heat: Heat remaining in the reactor circuit when the nuclear reaction has stopped.

Scientific notation: Scientific notation provides a convenient way of writing very large and very small numbers. For example, 2.2×10^4 is equivalent to 22,000, and 5.1×10^{-6} is the same as 0.0000051.

Separative Work Unit (SWU): This is a complex unit which is a function of the amount of uranium processed and the degree to which it is enriched, that is, the extent of increase in the concentration of the ^{235}U isotope relative to the remainder. The unit is strictly: Kilogram Separative Work Unit, and it measures the quantity of separative work (indicative of energy used in enrichment) when feed and product quantities are expressed in kilograms. For example, to produce one kilogram of uranium enriched to 3.5 per cent ^{235}U requires 4.3 SWU if the plant is operated at a tails assay of 0.30 per cent, or 4.8 SWU if the tails assay is 0.25 per cent (thereby requiring only 7.0 kg instead of 7.8 kg of natural U feed). About 100,000–120,000 SWU is required to enrich the annual fuel loading for a typical 1,000 MWe light-water reactor. Enrichment costs are related to electrical energy used. The gaseous diffusion process consumes some 2,400 kWh per SWU, while gas centrifuge plants require only about 60 kWh/SWU.

Sievert (Sv): Unit indicating the biological damage caused by radiation. One Joule of beta or gamma radiation absorbed per kilogram of tissue has 1 Sv of biological effect; 1 J/kg of alpha radiation has 20 Sv effect and 1 J/kg of neutrons has 10 Sv effect.

Source: Anything that may cause radiation exposure – such as by emitting ionising radiation or by releasing radioactive substances or radioactive materials – and can be treated as a single entity for protection and safety purposes.

Source term: The amount and isotopic composition of material released (or postulated to be released) from a particular facility.

Spent fuel: Fuel assemblies removed from a reactor after several years' use.

Stable: Incapable of spontaneous radioactive decay.

Steam generator: A large heat exchange through which superheated water at high pressure in the primary circuit of a reactor passes. The heat from the water is transferred to water passing through the secondary side of the steam generator at lower pressure, which is allowed to boil to produce steam to drive a turbine to generate electricity. Also known as a boiler.

Tailings: Ground rock remaining after particular ore minerals (e.g. uranium oxides) are extracted.

Tails: Depleted uranium (*cf.* enriched uranium) with about 0.3 per cent ^{235}U.

Thermal reactor: A reactor in which the fission chain reaction is sustained primarily by slow neutrons and hence requiring a moderator (as distinct from Fast Neutron Reactor).

Transient: Short-lived change in the power level of an operating reactor.

Transmutation: Changing atoms of one element into those of another by neutron bombardment, causing neutron capture.

Transuranic element: A very heavy element formed artificially by neutron capture and possibly subsequent beta decay(s). Has a higher atomic number than uranium (92). All are radioactive. Neptunium, plutonium, americium and curium are the best known.

Tritium (T): Tritium is the isotope of hydrogen with one proton and two neutrons, T = 3H. Tritium decays with a half-life of 12.3 years, and thus occurs in nature in concentrations too small to be recovered. Tritium, however, can be produced by neutron reactions with the lithium isotope ^{6}Li. Tritium is radioactive, decaying by beta decay to the helium isotope ^{3}He.

Uranium (U): A mildly radioactive element with two isotopes that are fissile (^{235}U and ^{233}U) and two that are fertile (^{238}U and ^{234}U). Uranium is the basic fuel of nuclear energy.

Uranium hexafluoride (UF_6): A compound of uranium which is a gas above 56 °C and is thus a suitable form in which to enrich the uranium.

Uranium oxide concentrate (U_3O_8): The mixture of uranium oxides produced after milling uranium ore from a mine. Sometimes loosely called yellowcake. It is khaki in colour and is usually represented by the empirical formula U_3O_8. Uranium is sold in this form.

Vitrification: The incorporation of high-level wastes into borosilicate glass, to make up about 14 per cent of it by mass. It is designed to immobilise radionuclides in an insoluble matrix ready for disposal.

VVER: Russian-designed pressurised-water reactor sometimes referred to as the WWER; abbreviated from 'water–water' reflecting its use of water as coolant and moderator.

Waste: High-level waste (HLW) is highly radioactive material arising from nuclear fission. It can be recovered from reprocessing spent fuel, though some countries regard spent fuel itself as HLW. It requires very careful handling, storage and disposal. Low-level waste (LLW) is mildly radioactive material usually disposed of by incineration and burial.

Yellowcake: Ammonium diuranate, the penultimate uranium compound in U_3O_8 production, but the form in which mine product was sold until about 1970. See also Uranium oxide concentrate.

Zircaloy: Zirconium alloy used as a tube to contain uranium oxide fuel pellets in a reactor fuel assembly.

Further reading

IAEA: *Management of life cycle and ageing at nuclear power plants: improved I&C maintenance*, IAEA-TECDOC-1402, Vienna, 2004

World Nuclear Industry Handbook (Nuclear Engineering International, Wilmington Publishing, 2005)

Semat, H., and Albright, J.R.: *Atomic and Nuclear Physics* (Chapman, 1983)

Aubrey, C.: *Meltdown: the collapse of the nuclear dream* (Collins & Brown, London, 1991)

Duncan, T.: *Advanced Physics: Fields, Waves and Atoms* (John Murray, London, 1981)

Radiation: doses, effects, risks, UN Environment Programme

Patterson, W.C.: *Going Critical* (Paladin, London, 1985)

Mosey, D.: *Reactor Accidents* (Nuclear Engineering International, Wilmington Publishing, 1990)

Chernobyl's Legacy: Health, Environmental and Socio-Economic Impacts, International Atomic Energy Agency, Vienna, 2005

Decarbonising the UK: energy for a climate conscious future, Tyndall Centre for Climate Change Research, 2005

Our Energy Future: creating a low carbon economy, Department of Trade & Industry, 2003

Managing the Nuclear Legacy, Fifth Report of the Trade & Industry Committee, HMSO, 2002, ISBN 0 21 500477 9

The Future of Nuclear Power: an interdisciplinary MIT study, Massachusetts Institute of Technology, 2003

Security of Energy Supply, Second Report of the Trade & Industry Committee, HMSO, 2002, ISBN 0 21 500155 9

National Atomic Museum: www.atomicmuseum.com

Uranium Information Centre: www.uic.com.au

World Nuclear Association: www.world-nuclear.org

World Association of Nuclear Operators: www.wano.org.uk

Nuclear Energy Institute: www.nei.org

US Nuclear Regulatory Commission: www.nrc.gov

Index